Movement and Connectivity

Movement and Connectivity

Configurations of Belonging

Edited by

JAN KETIL SIMONSEN

KJERSTI LARSEN

ADA I. ENGEBRIGTSEN

PETER LANG

Oxford • Bern • Berlin • Bruxelles • Frankfurt am Main • New York • Wien

Bibliographic information published by Die Deutsche Nationalbibliothek.
Die Deutsche Nationalbibliothek lists this publication in the Deutsche Nationalbibliografie; detailed bibliographic data is available on the Internet at http://dnb.d-nb.de.

A catalogue record for this book is available from the British Library.

Library of Congress Control Number: 2017945170

Cover design by Peter Lang Ltd.

ISBN 978-1-78707-550-4 (print) • ISBN 978-1-78707-551-1 (ePDF)
ISBN 978-1-78707-552-8 (ePub) • ISBN 978-1-78707-553-5 (mobi)

Published by Peter Lang Ltd, International Academic Publishers,
52 St Giles, Oxford, OX1 3LU, United Kingdom
oxford@peterlang.com, www.peterlang.com

This publication has been peer reviewed.

Contents

Acknowledgements

This book is a result of collaborations between researchers who carried out ethnographic field research in different parts of the world, and who share a common interest in migration and the relation between livelihood and shifting modes of social organization. The collaboration was motivated by the idea of approaching processes of migration and movement through the concept of 'mobility'. The aim has been to rethink our ethnographies within a conceptual framework of the anthropology of (im)mobility. Our method has been to bring our diverse and individually produced ethnographies on movement in time and space into dialogue with each other at a higher level of abstraction for the purpose of more systematically probing the theoretical implications of a turn to 'mobility'; re-questioning and re-thinking the relationship between space, time and livelihoods, and how differently motivated geographical movements are perceived as part of a wider social complexity. We thought this a viable way of unpacking and questioning the legitimacy of the tacit sedentarist assumptions of social research.

The project has been a collaboration between researchers from Norwegian Social Research (NOVA) at Oslo and Akershus University College of Applied Sciences; the Department of Ethnography, Numismatics and Classical Archaeology, Museum of Cultural History, at the University of Oslo; and the Department of Social Anthropology at Norwegian University of Science and Technology (NTNU) in Trondheim. Staff at these departments constitute the core members of the Norwegian Network for the Anthropology of Mobilities, supported by the Norwegian Research Council, Program for Institutional Development and Collaboration (grant no. 222825). Earlier drafts of the chapters were presented at the Biennial Trondheim Colloquium in Social Anthropology, organized by the Department at NTNU. We are grateful to Nigel Rapport, Professor of Anthropological and Philosophical Studies at the University of St Andrews, for generously writing an epilogue to the chapters.

Jan Ketil Simonsen, Kjersti Larsen, and Ada I. Engebrigtsen
Trondheim and Oslo, January 2017

KJERSTI LARSEN, JAN KETIL SIMONSEN
AND ADA I. ENGEBRIGTSEN

Introduction

This book discusses the question of mobility through time. Having carried out longitudinal fieldwork, the authors of the various chapters explore oscillation between movement and fixity through long-term ethnographies. Such ethnographies encompass life histories of individuals, household developments and the evolution of family and community configurations, and networks. The venture involves an analysis of how values, ideas and practices are affected by socio-political and economic change on the national and international levels and in what ways people's mobility-formation interacts with socio-cultural reproduction and change. A longitudinal perspective foregrounds how meanings and practices of mobility change in relation to political and economic regimes, in life-cycles of individuals and groups, and across generations. It allows an investigation of how mobility forms part of social and cultural processes constitutive of new social forms, networks and economic circumstances. In particular, a longitudinal perspective enables an understanding of interconnections between physical and social mobility – a connection which is emphasized throughout.

From this perspective, the authors discuss mobility within their regions of study, specifically Southern and Eastern Africa, Oceania and Europe, which in turn constitute the context for their individually produced empirical material in relation to the movement and connectivity of groups and persons within a specific time span, not only in terms of historical processes of change, but also from a subjective perspective. They broach questions such as: how does time interplay with feelings of belonging, with exclusion, uncertainty and expectation? and in what manner do imagined versions of and expectations for the future affect practices and experiences of the present? Thus, attention is paid to the various ways in which mobility is significant over time; in what manner it intersects with socio-political reform and reorientation, and the way the very meaning of movement may change with time and according to the social circumstances in which people find themselves. To this end, the diachronic dimension of the material is highly significant in the understanding of relationships between mobility, connectivity and configurations of belonging.

This book emphasizes not only movement in itself, but the manner in which a focus on mobility may advance an understanding of processes constitutive of society. As David Parkin (1998: ix) has previously argued: 'movement within one or two generations rather than fixed settlement has generally characterized human populations'; this is a dimension which many anthropologists have already incorporated into their analyses. Still, if modes of mobility are to be considered constitutive of society, social configurations and experiences of both connectivity and belonging have yet to be investigated. Connectivity, that is, the state of being related, does not presuppose a fixed settlement, but it does depend on the ability or wish to establish and maintain communication between people and localities. Following from this idea, belonging is perceived in terms of 'experience': a phenomenology of locality where contestable collective memory and ways of remembering would play a significant role in the creation of identity configurations' (see also Lovell 1998: 1). This means that 'belonging' is mainly addressed from the subjectivity of those who move, have moved or stayed put. In this volume, belonging will thus not be scrutinized according to its current usage within discourses on autochthony, as well as in current trends of identity-politics, where increasingly finer distinctions are drawn between 'us' and 'others' (Geschiere 2009: 92). The chapters rather apply longitudinal ethnographies to reinsert the dimensions of experience and social memory into the study of movement and migratory practices, that is, modes of mobility.

Comparatively, the various case studies show different social effects of mobility within and across different societies and historical periods. In Chapter 1, Øivind Fuglerud discusses how, in the case of Tamil Sri Lanka, international migration has transformed previous caste structures into an economically constituted class formation. Contrary to how international migration has transformed previous forms of social structures, Jennifer

Hays and Velina Ninkova (Chapter 4) highlight how movement within a designated territory remains crucial for the reproduction of Ju|'hoansi San identity and for the reproduction of sociality moored in nomadic values and ethos. While a mobility approach brings up the importance of territory among the Ju|'hoansi San, in Chapter 2 Ada I. Engebrigtsen discusses the socio-political and historical effects of de-territorialization and a self-image as mobile subjects in the case of Norwegian and Romanian Roma. Engebrigtsen conveys how a culturally formatted perception of movement and non-submission to state control generate a particular kind of life-world and sense of self. In Chapter 3, Kjersti Larsen explores identity configuration among residents of Zanzibar Town. She focuses on generational movements and the ways in which spatial emplacement beyond Zanzibar produces a particular form of 'multiplex' belonging. Larsen emphasizes the significance of the historical dimension to discern socio-cultural effects of oscillation between mobility and stasis through the life trajectory of individuals and groups. In a different manner, the relation between mobility and stasis is brought out by Jan Ketil Simonsen in his case study of the Mambwe-speaking people of Zambia (Chapter 6). Applying a mobility approach, he examines kinship relations as relations facilitating a sedentary life-world in a society where people constantly need to move between places. These two case studies equally show how a mobility approach captures the social fabrication of continuity. In a similar vein, in the context of migration between Tonga and New Zealand, Arne Aleksej Perminow (Chapter 5) investigates how Tongans in New Zealand continuously engage in a recreation of what they perceive to be a Tongan life-world.

Conjointly, the chapters bring to the fore the manner in which mobility practices and experiences within or against complex, transnational configurations of political, economic and socio-cultural relations that manage and control the (im)mobility of individuals and groups – what Glick Schiller and Salazar (2014) term *regimes of mobility* – in turn fashion people's perceptions of movement and the social relationships they involve: how are people's senses of belonging formed? In what ways do people mould and live connections through time and across localities, and what are enduring experiential and material interconnections composed of?

Anthropology and the question of mobility

In the discipline of anthropology, processes of social and cultural continuity and change have often been discussed in contexts of spatial movements, usually understood in terms of migration; either as labour migration (Mitchell 1956, Pottier 1988, Ferguson 1990a and b) or forced migration and refugeeism (Bascom 1998, Indra 1999, Black and Koser 1999). Moreover, studies of nomadic communities, in particular, have provided rich ethnographic material underlining how mobility becomes constitutive for society (e.g. Bohannan 1954, Lienhardt 1961 and 1966, Evans-Prichard 1969 and 1951). The analysis of the ethnographies were, however, fashioned according to the theoretical perspectives on social organization the respective researchers had in mind, and, until recently, there has been a tendency to see mobility more as a sign of societies and social systems in crises, than constitutive processes of society (Larsen 2003). Mobility of people within and between state boundaries, or even across continental boundaries, has mostly been approached by examining the problems created in the wake of movements either with regard to the society people are moving from or the societies they move to. Their movement is then seen as indicative of social and cultural disruption and disorder: 'acculturation' or the denial of modernization.

In this volume, the authors would like to avoid such an understanding as, in most cases, a conflation between mobility and disarray would remain intricately entangled in the ideologically constructed nature of the nation-state and its emphasis on closed social and cultural entities associated with territorial boundaries (Daley 2001).

The individual case studies presented here show precisely how mobility and its changing pattern and meaning through time shape the configuration of belonging and identities, which is pivotal to the constitution and reproduction of societies and communities. In the case of Sri Lanka, Fuglerud shows how Tamils settled in Norway situate their life within a context of what he terms 'transnationalism by proxy', indicating the limits of mobility and how Tamils recreate transnationalism not necessarily through physical movement, but by means of ideological commitment and remembrance, kinship networks and new communication technologies. Engebrigtsen's discussion of Roma as mobile subjects reveals how transnational movement has become almost imperative to the Roma's habitus; being 'on the move' and, perhaps, 'in-between' becomes a profound dimension of their identity and livelihood. In fact, it remains significant to their dignity as a de-territorialized community vis-à-vis the societies and state authorities between which they oscillate. Mobility not only produces unprecedented forms of coping strategies, but may also be decisive for certain livelihoods. Simonsen shows how movements of Mambwe-speaking people in Zambia between households in urban and rural areas, as well as within the urban setting, are motivated by socio-economic aspirations. Their access to and movement between households of relatives in pursuit of a better life is facilitated by a culturally shared notion of shared territorial belonging. In contrast, for the Roma discussed by Ada I. Engebrigsten, mobility is interconnected with them being dispossessed of a territorial basis. Thus, mobility for them becomes a tactic for economic and political coping.

In many cases, mobility remains a subsistence strategy. Yet, governments and their policymakers continue to consider sedentarization of transhumance and nomadic populations as a laudable development goal and, hence, as the kind of alternative livelihood that should be adaptable to and desirable for strengthening the vision and the hold of the nation-state (Wilson 1995, Harir 1994, Fratkin 1999, Manger 2001). Settlement and the reorganization of society that follows – of its practices, ideas and relationships – affect not only the system of production but also people's life-worlds. In Chapter 4, on Ju|'hoansi San of Namibia, Hays and Ninkova discuss how government- and NGO-initiated discourses on socio-economic development presuppose notions of sedentarism and vertical organization of society, while, locally, everything is perceived as more egalitarian and as always being in flux and in motion. It is interesting that in both this study of the nomadic Ju|'hoansi San of Nambia and their entanglement with development policies and Fuglerud's discussion of how Norwegian migration policies frame the life trajectories of Tamil women and men, it is clearly shown how abstract administrative categories of development, migration and social change fail to capture the complex relationship between movement and emplacement that a focus on actual practices of mobility bring forward. It seems that the more abstract and ideologically formatted categories of development and migration are still moored in the idea that domestication and sedentarization are the locus of progress – of livelihood security and community formation.

A mobility perspective brings out the dynamics between movement and stasis in terms of coping, as well as, local perceptions of connectivity and belonging. As early as 1992, Malkki questioned the 'sedentary metaphysics' inherent in social theory. And, since John Urry's influential *Sociology beyond societies: Mobilities in the 21st century* (2000), mobility has been a significant dimension in studies of contemporary social, political and economic processes. More than being an additional dimension in social analysis, a turn to mobility certainly challenges our understanding of society. It also questions the fundamentally territorial and sedentary basis of societies and cultures (Gupta and Ferguson 1992, Hannam, Sheller and Urry 2006). To a certain extent, mobility has become a buzzword for analysing the current global situation, often portrayed in terms of instability, super-diversity and transnational flows. Images of movement in all directions, followed by rapid changes and insecurities, as well as by formations of new stabilities, complete the depiction. Urry's main argument for discussing what he saw as new forms of sociality and social configurations was to challenge the rather bounded and static concept of society he found in sociological analysis.

With Urry, the mobile aspects of existence became explicit. He suggested that we should see the social as mobility rather than as society. Correspondingly, but on the level of the individual, Nigel Rapport, in the epilogue, universalizes mobility as fundamental to human existence. Rapport points out and discusses how general anthropological perspectives imply that movement is an individual human capacity intrinsic to the sense people make of the self and the world. He offers an alternative angle to the perspective sensitive to the social and historical dimensions characterizing the chapters of the book. The chapters argue how mobility and movement may be consequential for identity configurations and social forms alike. They focus on categories that account for continuities, beyond the immediate perception at different moments in time and space. In the case of the Norwegian Roma and the Romanian Romanies, Engebrigtsen examines the extent to which mobility is essential to Roma identity – a configuration that produces an opposition between them and the state in public negotiations of interests and claims. Hays and Ninkova centre the significance of understanding the disparities between a nomadic and a sedentary ethos, while Larsen, in her case study from Zanzibar, investigates the formation of multiplex identities in a society historically constituted through migration and adaptation. The cultural imagination of mobility, and the ways in which identity and belonging are configured in relation to mobility, are always relative to various forms of society.

Research following the 'mobility turn' has encompassed large-scale movements of people, objects and capital, as well as daily processes at a local level and the interconnections of these movements. Although large-scale movements have been approached in terms of migration for a long time, the concept of 'migration', as Urry points out (2000), is too limited to analyse the multi-dimensional aspects of the mobility, and, moreover, too bound up with one-directional and forced movements between bordered nation-states. Arguably, in spite of more or less forced or involuntary movements and transformations of livelihoods being part of the reality of mobility (Salazar and Smart 2012), the concept of 'mobility' may easily evoke the notion of freedom to move and the ability to change. While border crossing and migration in general were once seen as deviant exceptions to settled normality, today certain kinds of movement are 'being promoted as normality, and place attachment as a digression or resistance against globalizing forces' (Salazar and Smart 2012, ii). In Chapter 1, Fuglerud explores how, in the case of Tamil refugee-migrants to Norway, a transnational orientation towards 'homeland' over time has promoted social mobility within Norway. This orientation has produced a horizontal solidarity, anchored in a concept of national identity, rather than through their previous hierarchical caste distinction. The chapter clearly conveys how a perception of global movement has become vital to understanding national rootedness, and shows also how communities and people make sense of their fluctuation between movement and stasis when territorially based. Engebrigtsen's study of the Roma shows a similar idea of fluctuation, but in the context of Roma groups' historical and ideological de-territorialization. The shift towards mobility in understanding and valuing must be understood in relation to the geo-political changes that have taken place: from the nation-state as a normative and empirical reference for governance to the present-day globalized political reality, with the increasing permeability of national borders (Wimmer and Glick Schiller 2002). These transfigurations suggest how various forms of mobility prove essential to an understanding of societies (Clifford 1992), as the term 'mobility' itself refers to all kinds of movement, such as pastoralism, nomadism, trade, travel, migration and refugeeism (de Bruijn et al. 2001). A focus on mobility as a normal aspect of life contrasts more hegemonic understandings that emphasize the importance of staying in one('s) place both with regard to identity formation and socio-cultural continuity. The chapters in this book therefore explore mobility as a characteristic and genuine feature of certain societies, a feature inherent in processes through which social organization and societal continuity become reproduced. For the Mambwe in Zambia, discussed by Simonsen (Chapter 6), rootedness in the land seems crucial for their shared identity, which in turn facilitates their spatial movement between kin in search of social mobility in places away from their land. At the same time, their reciprocal movement between kin-based households is precisely what reproduces their particular form of social organization and diversified subsistence strategy. Hence, the propensity for mobility is, in certain ethnographic contexts, crucial for the reproduction of a particular form of social organization, involving both those who move and the accommodating communities.

Mobility from below

Notably, all chapters in this volume mainly discuss mobility 'from below', foregrounding relations between spatial and social mobility as processes constitutive of society, but also of identity – processes through which mobility is perceived and experienced as part of an expected flow of life. Attention is given to local understandings: how do people see their own local life-world and its (un)connectedness to other societies? Such an approach presupposes, precisely, that research should be grounded in what Maria Mies (1983) in her methodology defined as 'a view from below', inspired by a concern about how to grasp local understandings and perspectives. This does not imply that a retreat to relativism is necessary in order to understand the reasons and consequences of various concepts and forms of mobility. It rather suggests that we engage in comparison debating the meaning of mobility with our 'others'. Despite the many claims that the world of today is one of globalized processes equally shaping life as well as its meaning for people in all of the various centres and peripheries, it is here considered that forms of movement should be approached according to how they relate to the constitution of various forms of societies and to the different ways of being in and relating to life situations. Apart from the mobility perspective suggested by Rapport in this volume's epilogue, which scrutinizes mobility as part of individuals' physical and psychological capacities for bodily movement, behaviour and action, the chapters suggest how an approach sensible to history may fashion mobility and social structure in different societies. Engaging in this effort, it should be kept in mind that what are here referred to as local understandings will always be informed and influenced by a multitude of different experiences, rumours, impressions and knowledge about what goes on elsewhere and why. And, inferring from this, how do imaginations of the future, of the 'not yet' formulated by Barbara Adams and Chris Groves (2007), affect experiences and practices in the present? In Chapter 3, Larsen investigates how ideas about potential spatial movement nourish people's future prospects in Zanzibar. Despite current limitations to both spatial and social mobility in a society which has historically been forged through complex interconnections across the Indian Ocean, people remain connected transnationally. Discourses on travelling and migration, as well as about places and lives elsewhere, are still integral aspects of their present and local livelihoods. To some extent, this confirms that local understandings, far from being narrow, or even uninformed, reveal how external issues, events and debates, over time, are interrelated and experienced, perceived and adjusted to particular places by particular people managing relationships within, as well as beyond, spatial and ideological boundaries.

Mobility through time, transformative processes and societal continuity

Turning to time in approaching human mobility suggests perspectives and methods sensitive to the fact that human lives unfold over time and that human mobility incorporates temporal as well as spatial dimensions. For instance, Sheller and Urry (2006) have encouraged the study of how place and movement are conceptualized by both sedentary and mobile subjects and groups, while the meaning of place and place-making for mobile subjects and the manner in which perceptions of home and movement influence each other over time, is another recent field of study (Rapport and Dawson 1998, Williksen and Rapport 2010). Importantly, the perception and meaning of places are also reworked over time. The strengths of anthropological methodology are precisely its interest in and capacity to discover the time perspectives within which social agents themselves frame their acts, lives and selves (Hirsch and Stewart 2005). In Chapter 5, Perminow discusses the prospects of a Tongan form of sociality moored in perceptions of 'mutual flows and nurturing spaces' in a situation in which a number of people leave the island, as well as the local practices, behind. Perminow shows how the force of cultural notions of 'homemaking' and how norms of movement and flow in nurturing spaces in order to (re-)produce particular places continue to impinge on Tongans' life-world wherever they happen to assemble. He discusses the extent to which a mobility approach captures a sense of continuity in a situation in which people move into another society, which may not eventually accommodate the values those who migrate perceive as decisive to a Tongan life-world. This theme links up with Hays and Ninkova's discussion (Chapter 4) of how the historically configured nomadic ethos of the Ju|'hoansi San is closely interconnected with the perpetuation of a particular form of livelihood.

Linking mobility and historicity is not simply concerned with different forms of human travelling, but rather with how motion becomes culturally and socially meaningful, and how these may also cause disruptions in established relational and identitary processes, or, alternatively, how more lasting social and cultural forms may be reproduced through disruption and change. Such processes are captured by both Simonsen's study of the Mambwe in Zambia and by Perminow's study of Tongans in Tonga and New Zealand. Both case studies investigate to what extent a mobility approach may tease out the way abstract kinship statuses become actualized and re-appropriated in situations of circular, repetitive forms of moving between places and households, including transnational migration.

The relevance of a focus on temporality in relation to mobility is that it allows us to first understand and subsequently analyse how putative demarcations between 'places' and 'spaces' permeate in different ways, ways that challenge the very idea of society and culture as predominantly sedentarized and bounded. Turning to time in exploring human mobility, the chapters of the volume are attentive to historicities of mobility as these are continually produced by people who move and/or have moved in different parts of the world and within ideological and societal contexts that are differently constituted. The authors approach historicities of mobility, engaging ethnographically diverse investigations of the modes in which people on the move actualize the past in the present with the future in mind.

In particular, the different contributors examine mobility by connecting it to local as well as to extra-local historical events that have impacted on already existing modes of movement and, in turn, on interconnectivity. All the chapters in this volume are attentive to how the involved life-worlds relate to and are impacted by wider socio-political contexts and their economic realities. They investigate the ways in which migratory practices or more or less continuous relocations are engraved in people's experiences and everyday-life practices, as well as in the social memory of their communities. In Chapter 6, which focuses on the Mambwe-speaking people in Zambia, Simonsen explores how households dispersed across wide geographical space constitute nodes in a network of relatives and kin and represent temporary moorings for persons on the move. Studying these interconnections, a mobility approach makes possible a discussion of how actual practices of mobility, not only abstract kinship categories, become constitutive of Mambwe society, as well as facilitating a diversified livelihood. The genealogical connections between hosts and travellers may be close or remote; yet they are always flexible and accommodating. Mobility depends on the recognition of hospitality towards kin and a sense of connectivity provided by the Mambwe conceptualization of kinship, which equally ignores time and space and is marked by an emphasis on 'wealth-in-people'.

Recapturing how modes of mobility through time, their openings and limitations, are inherent in the reproduction of society, if not of state politics, the authors analyse ethnographic material collected over the last twenty to thirty years. Glick Schiller and Salazar (2014) have suggested the term *regimes of mobility* to focus on the inherent power dimensions in mobile lives and mobile thought, and to investigate the binary logic of movement and stasis. 'Regimes' denotes both global and regional forms of capital accumulation and individual states' policies to include, exclude, and regulate movement inside its borders and changing regulatory structures and surveillance concerning movement between nation-states – as well as economic conditions enabling or forcing subjects to stay or to go. All these regimes and facets fashion individual and collective movement and create new forms of mobility in terms of 'illegality' and criminality (see Chapter 2). Hence, historically, what are considered 'regimes of mobility' have changed according to ongoing national, international and global politico-economic processes. As indicated above, a distinct feature of the book is that longitudinal ethnographies enable the authors to examine local and experience-based understandings of mobility, including modifications in their own social and ideological rationale. Thus, in this book, mobility becomes an 'experience-near' concept through which the authors analyse how people on the move or those made to stay put perceive social relationships and the horizon of their life-worlds. Consequently, attention is paid to people's experiences of being on the move, the social relations they involve and the way these become recomposed or, perhaps, reshuffled in the wake of changing political and economic circumstances. In what ways do people's or communities' preconfigured mental and physical readiness for engaging in migration, their perceptual practices, actually fashion their understandings of the relationship between movement, fixity, and connectivity?

Importantly, the book questions the manner in which changing circumstances facilitate and restrict socially and culturally engraved modes of mobility and fixity for different groups or categories of populations within, as well as across, national boundaries. In different ways, the chapters all acknowledge how spatial and social mobility are often interrelated and connected with shifting formations of identity. Economically and status-wise, spatial mobility may bring or even accelerate either downwards or upwards social mobility, while in other situations people aspire for spatial mobility and can only imagine various forms of social mobility that could ideally result from a change of both physical place and ideological space (see Chapter 1, Chapter 3, and Chapter 6). In many situations, actual and imagined forms of social mobility interlink with the aspiration to eventually return to a particular place of departure, and/or enter into a livelihood grounded in a remittance-based economy. The complex economic relations that may be embedded in a term like 'remittance', are elucidated, for instance, in Fuglerud and Perminow's respective contributions. The difference between the two economies evokes reflections on how, for instance, understandings of providence, maintenance and investment are always relative to the historical and cultural context in which they occur. In order to understand historical, contemporary and emerging forms of social organization, an exploration of the relationship between time, spatial and social mobility is, indeed, crucial. The link, for instance, between social mobility and migration has been widely researched in sociology, anthropology and other disciplines. Yet the way different practices of spatial mobilities interlink with different prospects for social mobility (upwards and downwards) has not yet been properly researched to any systematic and comparative extent. However, what is empirically well established is that what often motivates physical movement inside, as well as across, national borders is an expectation of a better life for oneself and one's family, most significantly for one's own children. Wars, deportations, persecution, trafficking, and ecological breakdowns represent meaningful factors initiating migration. They often, yet not automatically, may lead to a downward mobility, at least for the first generation in a new place.

By putting diverse and individually produced ethnographies on movement in time and space into dialogue with each other at a higher level of abstraction, our wish has been to re-direct and re-think relationship between space, time and livelihoods, and how differently motivated geographical movements may be perceived and lived as part of wider social complexities. The perspective advocated by Rapport in the epilogue may at a first glance stand out, not only as an alternative perspective, but as a counterpoint to the mobility approach argued by the other case studies in this volume. While the authors of the chapters emphasize the significance of history and the local social contexts, Rapport side-tracks from the socio-cultural context foregrounded by the authors and develops a perspective on mobility as a capability intrinsic to human existence. In fact, Rapport's perspective highlights what has been the aim of this book, namely, to explore the ways in which a mobility approach, including both an historical dimension and a view from below, advances the understanding of the complex relationship between different migratory practices, experiences of belonging and the kinds of movement and connectivity that make and re-make people, as well as their societies.

Bibliography

Adam, Barbara, and Chris Groves. 2007. *Future Matters. Action, Knowledge, Ethics.* Leiden and Boston: Brill.

Bascom, Johnathan. 1998. *Loosing Place: Refugee Populations and Rural Transformation in East Africa.* New York: Berghahn Books.

Black, Richard and Khalid Koser (eds). 1999. *The end of the Refugee Cycle?: Refugee Repatriation and Reconstruction (No. 4).* New York and Oxford: Berghahn Books.

Bohannan, Paul. 1954. 'The Migration and Expansion of the Tiv'. *Africa*, XX (VI): 2–16.

Brenner, Neil. 2004. *New State Spaces: Urban Governance and the Rescaling of Statehood.* Oxford: Oxford University Press.

Clifford, James. 1992. 'Traveling Culture'. In *Cultural Studies*, edited by Lawrence Grossberg, Cary Nelson, and Paula Treichler, 96–116. New York: Routledge.

Daley, Patricia. 2001. 'Population Displacement and the Humanitarian Aid Regime: The Experience of Refugees in East Africa'. In *Mobile Africa*, edited by Mirjam de Bruijn, Rijk van Dijk and Dijk Fovken, 195–212. Leiden: Brill.

Evans-Prichard, E. E. 1951. *Kinship and Marriage Among the Nuer.* Oxford: Clarendon Press.

——1969. *The Nuer: A Description of the Modes of Livelihood and Political Institutions of a Nilotic People.* New York: Oxford University Press.

Ferguson, James. 1990a. 'Mobile Workers, Modernist Narratives: a Critique of the Historiography of Transition on the Zambian Copperbelt [Part one]'. *Journal of Southern African Studies*, 16 (3): 385–412.

——1990b. 'Mobile Workers, Modernist Narratives: A Critique of the Historiography of Transition on the Zambian Copperbelt [Part two]'. *Journal of Southern African Studies*, 16 (4): 603–21.

Fratkin, Elliot. 1999. 'East African Pastoralism in Transition: Masaai, Boran, and Rendille Cases'. *African Studies Review*, 3 (44): 1–25.

Geschiere, Peter. 2009. *The Perils of Belonging. Autochthony, Citizenship, and Exclusion in Africa & Europe*. Chicago and London: The University of Chicago Press.

Glick Schiller, Nina, and Noel. B. Salazar. 2013. 'Regimes of Mobility Across the Globe'. *Journal of Ethnic and Migration Studies*, 39 (2): 183–200.

Gupta, Akhil, and James Ferguson. 1992. '"Beyond 'Culture"': Space, Identity, and the Politics of Difference'. *Cultural Anthropology*, 7 (1): 6–23.

Hannam, Kevin, Mimi Sheller, and John Urry. 2006. 'Editorial: Mobilities, Immobilities and Moorings'. *Mobilities*, 1 (1): 1–22.

Harir, Sharif. 1994. 'Recycling the Past in the Sudan. An Overview of Political Decay'. In *Short-Cut to Decay: The Case of the Sudan*, edited by Sharif Harir and Terje Tvedt. Uppsala: Scandinavian Institute of African Studies.

Hirsch, Eric, and Charles Stewart (eds). 2005. 'Ethnographies of Historicity'. *History and Anthropology*, 16: 261–74.

Indra, Doreen Marie (ed.). 1999. *Engendering Forced Migration*. Oxford: Berghahn.

Larsen, Kjersti. 2003. 'Mobility, identity and belonging: The case of the Hawawir, Northern Sudan'. In *The Maze of Displacement. Conflict, Migration and Change*, edited by Nadarajah Shanmugaratnam, Ragnhild Lund, and Kristi Anne Stölen, 106–27. Oslo: Cappelen Damm Høyskoleforlaget.

Lienhardt, Godfrey. 1961. *Divinity and Experience: The Religion of the Dinka*. Oxford: Oxford University Press.

——1966. *Social Anthropology*. London: Oxford University Press.

Lovell, Nadia. 1998. *Locality and Belonging*. London and New York: Routledge.

Malkki, Liisa. 1992. 'National Geographic: The Rooting of Peoples and the Territorialization of National Identity Among Scholars and Refugees'. *Cultural Anthropology*, 7 (1): 24–44.

Manger, Leif. 2001. 'Pastoralist-State Relationships Among the Hadendowa Beja of Eastern Sudan'. *Nomadic Peoples*, 2 (5): 21–48.

Mies, Maria. 1983. 'Towards a Methodology for Feminist Research'. In *Theories of Women's Studies*, edited by Gloria Bowles, and Renate Klein, 117–39. London: Routledge & Kegan Paul.

Mitchell, James Clyde. 1959. *The Kalela Dance: Aspects of Social Relationships Among Urban Africans in Northern Rhodesia.* Manchester: Published on behalf of the Rhodes-Livingstone Institute by the Manchester University Press.

Parkin, David. 1998. 'Foreword'. In *Locality and Belonging* edited by Nadia Lovell, ix–xiv. London and New York: Routledge.

Pottier, Johan. 1988. *Migrants no More: Settlement and Survival in Mambwe Villages.* Manchester: Manchester University Press.

Rapport, N., and A. Dawson (eds). 1998. *Migrants of Identity: Perceptions of Home in a World of Movement.* Oxford: Berg.

Salazar, Noel. B., and Alan Smart. 2011. 'Anthropological Takes on (Im) Mobility'. *Identities*, 18 (6): i–ix.

Sheller, Mimi and John. 2006 'The new Mobilities Paradigm'. *Environment and Planning A*, 38 (2): 207–26.

Urry, John. 2000. *Sociology Beyond Societies: Mobilities for the Twenty-First Century.* London: Routledge.

Williksen, Solrun, and Nigel Rapport (eds). 2010. *Reveries of Home: Nostalgia, Authenticity and the Performance of Place.* Newcastle upon Tyne: Cambridge Scholars.

Wilson, Wendy. 1995. 'The Fulani Model of Sustainable Agriculture: Situating Fulbe Nomadism in a Systemic View of Pastoralism and Farming'. *Nomadic Peoples*, 36 (37): 35–52.

Wimmer, Andreas, and Nina Glick Schiller. (2002). 'Methodological Nationalism and Beyond: Nation-state Building, Migration and the Social Sciences'. *Global Networks*, 2 (4): 301–34.

ØIVIND FUGLERUD

1 Migration From Eelam (Sri Lanka): Terrorists, Model Citizens, and the People Left Behind

In this chapter, the relationship between the now well-established concepts of 'transnationalism' and 'diaspora' on the one hand and the more recent concept 'regimes of mobility' on the other is explored. I suggest that the latter facilitates explorations of disparities that are easily neglected when relying on the former two. I am interested in how these disparities play out over time; how people's physical movement has in the *longue durée* caused social effects that were not easy to foresee. One of these consequences is the creation of new bonds of solidarity among migrants, and, through this, a reinforcement of inequalities between those who left and those who remained behind. I will present the case of refugee-migration from Tamil Eelam (Sri Lanka) and explore how and to what extent Tamil emigration during the thirty-year civil war in Sri Lanka – normally conceptualized as 'forced migration' – fits into a social imaginary upheld by the dominant section of Tamil society. I use the term 'social imaginary' in the sense this term has been proposed by Taylor (2002), namely as

> the ways in which people imagine their social existence, how they fit together with others, how things go on between them and their fellows, the expectations that are normally met, and the deeper normative notions and images that underlie these expectations. (106)

The imaginary relevant to the present case portrays the hierarchical organization of Tamil society as being one of its essential qualities.

I will limit the exploration in this chapter to two topics: 1) the relationship over time between refugees' inclusion and social mobility in a country of settlement, Norway, and their transnational engagement and emotional bonds to their country of origin, and 2) the relationships between people who have emigrated as refugees and the society they have left behind. Regarding both issues there is a need to look at the ways in which networks are structured to assess the outcome of transnational processes. For this reason, I draw on notions of social capital in the discussion below. Let me begin with a short reference to analytical terms.[1]

Regimes of mobility

Glick Schiller and Salazar (2013) offer a careful critique of the recent 'mobility-turn' in studies of migration and studies of globalization more generally, in which all categories of people on the move, whether tourists, exchange students, labour migrants, or others tend to be studied through a common analytical lens. In staking out a direction for future research on mobility, they suggest six areas in need of further exploration, four of which I will explore: 1) the relationship between mobility and immobility; 2) the definition and meaning of mobility and immobility within relationships of power; 3) the development of a mobility approach that addresses the range of actors – including states – that define categories such as 'legal 'undocumented etc. and shape our conceptualization of mobility and stasis;

1 The chapter draws on several sources. The most important is my continuous contact with Sri Lanka and with Eelam Tamil communities in different parts of the world for the last twenty-five years. As part of this I have over the years been involved in several more specific processes of data collection, one set data of data drawn upon here are interviews with parents and youngsters of Tamil origin in Oslo conducted by my colleague Engebrigtsen and myself over a three-year period in 2005–7, some of which have been published earlier (e.g. Engebrigtsen and Fuglerud 2007; also Fuglerud and Engebrigtsen 2006). In addition, I refer to statistical data from a large survey named *Young in Oslo*, conducted by the research institute Norwegian Social Research in 2006, where Engebrigtsen and I both worked at the time, covering all students in Oslo in Grades 9, 10, and 11 (Øia 2007), and to official statistics collected by Statistics Norway. Finally, I draw upon media clippings and relevant secondary sources to support my arguments.

and 4) the development of approaches that expand our understanding of class privilege in travelling (Glick Schiller and Salazar 2013).

I admit to mixed feelings regarding this and similar (e.g. Faist 2013) discussions of the mobility-turn in the study of migration. Coming to the general discussion of mobility from the corner of the research field engaged with forced migration, refugee policy, long-distance nationalism, and diaspora-formations, and from the blood-soaked terrain of the Sri Lankan civil war (Fuglerud 1999, 2001, 2002, 2003, 2004a, 2009, 2011), it seems obvious that power, states, and legal definitions must be included in our analyses. At the same time, however, even within that corner of the field the growth in research interest and the enthusiasm with which concepts like 'transnationalism' and 'diaspora' were received since their launch in the early 1990s (Glick Schiller, Basch and Szanton Blanc 1992; Safran 1991) have led to a parallel value-degradation of these concepts. Too little effort has gone into distinguishing between different forms and varieties of diasporic identifications and the more precise character of transnational loyalties and dynamics. As noted by Brubaker (2005, 3), the concept of diaspora is in danger of losing its analytical punch, being applied to 'any and every nameable population category that is to some extent dispersed in space'. Sometimes excitement with the concepts slides into circular argumentation: Immigrants are seen as having transnational connections and because they have transnational connections they resist integration in countries of settlement. The thrill over new non-sedentary perspectives (cf. Wimmer and Glick Schiller 2002) has become part of the problem rather than part of the solution.

My wish here is to approach the discussion raised by Glick Schiller and Salazar (2013) about the need for one or more lenses when observing mobility through the way this question manifests itself empirically within the field of forced migration and diaspora studies. More specifically, I argue that using separate lenses for refugees, tourists, and other categories of migrants is not sufficient; these categories are themselves internally heterogeneous and cases must, therefore, be analysed concretely and individually. I see a dependence on administrative policy concepts – such as 'asylum-seeker' or 'forced migration' – as a problem for the field of refugee studies and something that continues to limit its scope (Bakewell 2008;

Black 2001; Landau 2007; Turton 2005). As noted by Turton (2005), 'paradoxically, the academic study of human displacement is less likely to be 'relevant' to policy the more closely it follows policy related categories and concepts in defining its subject matter and setting its research priorities' (277). As a field of research, forced migration studies is not a theoretically or conceptually delineated research area; at most it is a focus on a set of empirical issues and questions. Whoever the migrants are, we need to study each case concretely regarding possibilities of generalization. In so doing, the ideas put forward by Glick Schiller and Salazar are helpful. I will in the following discussion touch on four key ideas.

Transformation of the Tamil refugee

Glick Schiller and Salazar (2013) suggest that studies of mobility should research migration through an approach that does not use nation-states as units of analysis but 'studies the movement of people across space in relationship to forces that structure political economy. These forces include, but are not confined to, states and their policies' (10). Also, national and international policies should be considered within the same analytical framework (Schiller and Salazar).

This admonition is highly relevant to the case of Tamil refugees in western countries. Fleeing from the exploding violence of the civil war in Sri Lanka Tamil refugees started arriving in large numbers in the late 1980s and early 1990s and soon became a headache to European governments as well as to the UNHCR. Cels noted in 1989 that 'the debate about the status of Tamil asylum-seekers lies at the core of the problem of de facto refugees in Europe' (Cels, 1989). Immediately after the so-called 'Colombo riots' in 1983 – a euphemism for large-scale massacres of civilian Tamil-speaking citizens in the Sri Lankan capital – some countries took a relatively generous position on Tamil asylum-seekers. However, this group of refugees soon became a vehicle for increasing immigration control and travel restrictions. Hard-line policies imposed by some countries deflected the flow in other directions. For example, when the Dutch government imposed restrictive measures in 1985, about 2,000 Tamils arrived in Britain in a matter of weeks (Cels 1989). When the British imposed visa regulations in 1986, the flow was deflected to Switzerland. Similarly, when Denmark implemented new and stricter immigration regulations in 1986, Norway received its highest number of Tamil asylum-seekers ever in 1987. In all countries, the government response to the arrival of Tamils was accompanied by media coverage and public controversy.

No doubt the controversies partly had to do with numbers. Between 1986 and 1992, approximately 120,000 Tamils applied for asylum in Western Europe, a considerable figure even by today's standard. However, there was also another factor, namely the common opinion that the large majority of Tamils fleeing Sri Lanka in their individual capacities did not fulfil the criteria of the 1951 Convention because they had not been politically active in their country of origin and were not individually persecuted. To see the importance of this at the time one should realize that 'refugee' is not only a legal but also an ideological concept (Fuglerud 1997). As a European discursive formation, 'refugee' in the Cold War period from the late 1940s until the late 1980s incorporated two somewhat contradictory elements – the pathology of being uprooted and the longing for individual liberty, seen as universal to mankind. The idea of the person leaving everything behind and denouncing her or his original nationality to escape (communist) repression has been central to the refugee as an essentialized figure in western political mythology. Such a view is reflected in the 1951 Convention in which the individual is constituted as a social person not through family or cultural identity, but through her or his direct relationship to a state. The ideal and iconic refugee in this era, one may say, would be a person jumping the Berlin wall on her or his own, shifting state loyalty and social personality by this single action, denouncing communist repression and embracing western values in the act. The fact that Tamil refugee-migration followed a pattern of chain migration, and often involved planned violation of national law put their loyalty to settlement countries into question and encouraged the use of asylum procedures against them as a means of protecting the moral communities of western nation-states (Fuglerud 2004b). It indicated that loyalty to their social group was more important to them than their obligations as potential citizens in a new country. The importance of this dynamics to the handling of asylum applications became clear to me while working as an advisor to the Directorate of Immigration in the early 1990s. In internal memos assistance to migrating Tamil refugees from family members already settled in the west, and the migrants' use of professional, extra-legal travel agents to cross national borders illegally, were seen as documenting their ineligibility for asylum. Among officers of the immigration service, the common feeling was that if people wanted protection from the Norwegian government, they should at least follow prescribed rules and seek out the Norwegian embassy in their country of origin to apply for this in a proper way. The fact that they would never have been given permission to enter Norway if they did was never part of this discourse.

The understanding of Tamil refugees as somehow subversive also leaked into public space in the early 1990s, resulting in headlines like the following, taken from *Verdens Gang*, the largest national newspaper in Norway, all of them relating to the Tamil community:

> Mafia-suitcase with Norwegian Visas – Smuggler Arrested in Russia (*Verdens Gang*, 14 February 1992)
> Human Smuggling without Limits (*Verdens Gang*, 26 April 1992)
> Investigating the Tamil Community (*Verdens Gang*, 8 May 1992)
> Admits to Smuggling Humans (*Verdens Gang*, 4 September 1992)

The media stories of law-breaking came with a twist: while immigration authorities were concerned that Tamil asylum-seekers had no political involvement, and thus did not fulfil the criteria for being considered refugees, media could not resist spicing up their stories with the exotica of Tamil suicide bombers and brutal massacres. The following is a typical excerpt from that time:

> Under the cover of 'sleeping' Norwegian authorities, an extensive organization has been established in Norway with the aim of smuggling of humans, blackmail/death threats and forgery of travel documents. (*Verdens Gang*, 26 April 1992)

The organization in question was the Liberation Tigers of Tamil Eelam, LTTE, the Sri Lankan government's armed opponent in the civil war.

However, from the early 2000s, one could observe a remarkable transformation in the public image of this group of refugees in Norway. It was as though having finally curbed the inflow of new asylum-seekers through restrictive measures – including the deportation of many applicants in the late 1990s – the Tamils were ready for promotion in the ethnic hierarchy in Norway. Perhaps because of economic liberalism and new public management ideas, a tendency to focus on differences between the performance of immigrant groups with respect to work, welfare, crime, etc., evolved in the new century. In these debates the former 'Tamil terrorists' surfaced as model citizens of minority background, setting the standard for the evaluation of other groups of immigrants. One article in the national newspaper *Dagsavisen* reads, for example, 'The refugees from Sri Lanka are the most industrious in the Norwegian labour market. Tamils work more than Norwegians, and have twice the work participation of Somalis and Iraqis' (*Dagsavisen*, 16 March 2002). The story is illustrated with a picture of two Tamil gentlemen, Paul and Panch. The article continues,

> Dagsavisen meets Paul and his colleagues in the company Central Cleaning … There, all the 30 cleaning assistants are Tamils. Several of them have more than one job. 'I work here every evening, in addition I am a janitor. During the week-ends I am a bus-driver,' says Paul who was educated as a carpenter in Sri Lanka. 'I don't see much of the children, but it is for them that I work so much. I want them not to have to work in cleaning,' says Paul.

In this way, the new public image of the Tamil minority in Norway was casted into a well-known narrative about migration – the narrative of social mobility through hard work. This narrative is accompanied by stories of responsibility for one's own community. In particular, the weekend schools run by the Tamil community have received their fair share of attention. So also in the story in *Dagsavisen* (16 March 2002):

> The corridors and classrooms at the Tamil Resource and Counselling Centre are crowded with children and parents. The smells of fresh waffles and moth-balls make their way between duffel-coats and shining dresses with gold broidery. When the doors open at nine, children, youth and adults hurry to cultural activities and native language classes. Every Saturday throughout the year … All look forward to the weekend.

This portrayal is not an isolated instance; the notion of Tamil super-immigrants taking care of their own business has been promoted in all media and repeated again and again over the years. 'When politicians turn their back on native language education immigrants themselves establish language schools. Every weekend more than 200 children take their place at the Tamil Resource and Counselling Centre' one newspaper reads. 'The children receive extra tuition from Tamil university students' (*Aftenposten*, 28 December 2013). And in another report, under the title 'They are children of immigrants and doing best at school':

> Youngsters sweat over their homework. Others practice Indian song and music. Tamil Resource and Counselling Centre is the meeting place for young and old Tamils at Rommen in Oslo. They spread out over 1600 square meters and 18 rooms on two floors.
>
> Our vision is to be the foremost organization working to bring Tamils in Norway together around their common identity, says Rajah Balasingham, daily leader of the centre.
>
> There are no Tamil gangs in the streets of Oslo. We prevent psychological problems, our young educate themselves, we have a high participation in the labour market. (*Aftenposten*, 23 February 2010)

What has been realized neither by the media nor by the public in general is that the main factors behind what is here read as 'integration' are the same social characteristics that were seen as negative during the asylum period. Tamil refugees arriving in the 1980s and 1990s benefited greatly from being linked through family, caste, or village to the few early work migrants having arrived in the 'pre-discrimination period', some of whom even securing good positions in Norwegian society. For example, one of the earliest Tamil work migrants arriving in the 1970s worked his way up in the fisheries sector in northern Norway. In the 1980s, he recruited large numbers of newly arrived asylum-seekers of similar background for work in local fish procession factories (Fuglerud 1994). Another key figure, a female, unmarried nurse from a small village in Jaffna, Sri Lanka's northern Tamil-dominated peninsula, who settled in Oslo in the early 1980s, became instrumental in financing the travel to Norway of her extended family when the war escalated.

Further, the conscious reproduction of a diasporic transnational identity is a significant factor behind the success of this refugee group in Norway. The aforementioned weekend schools, providing additional school training and cultural identity to the generation growing up in Norway, have been important in this respect. Started up in a small way about twenty years ago, the network today has more than 2,000 members and runs schools in eighteen different locations. What is missing in the newspaper stories is that these schools are run by an organization that until the end of the war in 2009 was loyal to the LTTE and today is still grounded in a vision of Tamil national liberation.

Integration and transnational engagements

The perception of the Tamils in Norway as a model minority is not without empirical basis. To mention only a few statistical facts, a comprehensive comparative study of eighteen immigration-based minorities conducted by Statistics Norway in 2007, the latest of its kind, found that family income in families of Sri Lankan origin was well above the immigrant average. Moreover, 70 per cent lived in homes owned by themselves, and the second-generation educational pattern is equal to the one found in the general population (SSB 2007).[2] The percentage of homeowners in the study was based on numbers from 2001, and has since risen to more than 85 per cent (SSB 2008). The 2007 study concludes that Sri Lankans are 'relatively well integrated in the Norwegian society, and in many ways, they are more alike to the population in general than to other non-western immigrants' (SSB 2007, 138).

Engebrigtsen and I have elsewhere argued (Fuglerud and Engebrigtsen 2006, Engebrigtsen and Fuglerud 2009) that one important factor

2 In Norway, official statistics do not separate between ethnic groups. However, it is common knowledge that the clear majority of people originating in Sri Lanka living in Norway are Tamil-speaking.

explaining the success of Tamil immigrants in Norway is the social capital generated through what in transnational studies is now often termed 'homeland-oriented engagements'. Our argument, put bluntly, has been that the shared Tamil identity, established and enforced by the liberation movement and reproduced through various networks and meeting places managed by representatives and supporters of the LTTE in Norway, has helped their adjustment to Norwegian society. This may seem contra-intuitive. One would think that engagement in homeland-oriented activities would take focus away from local engagements and leave less time for community building in Norway. However, interviewing Tamil immigrants in 2005–7 about their detailed contacts with Sri Lanka, Engebrigtsen and I found that many interviewees had never been back to their country of origin and had rather sporadic contact with remaining relatives. Almost no one was involved in any business or religious activity with links to Sri Lanka. Instead, what we found was what may be called a 'transnationalism of the mind' or perhaps 'transnationalism by proxy', that is, people keeping a focus on the country of origin without themselves taking part in activities stretching beyond the local setting. In Norway, like in other western countries, radio-stations, meeting-places, and cultural events, giving people a sense that a Tamil community exists, have been initiated mainly by organizations loyal to the LTTE, particularly by the Tamil Co-ordination Committee (TCC). Organizations like the TCC, the later Transnational Government of Tamil Eelam (TGTE), and the Global Tamil Forum are transnational structures in a true sense. However, in Norway at least, they are run by a very limited number of dedicated cadres. The reasons common members have for supporting them economically or using the services they offer may be many and varied. Most important to many is the possibility they offer of joining, in the local setting, what may be called an imagined global Tamil community of suffering. Levitt and Glick Schiller (2004) have pointed to the importance of being involved in transnational activities without linking it with notions of identity. In the present case I would rather stress the possibility of expressing and sharing a homeland identity without having to be involved in actual transnational activities.

Networks and social capital

One important aspect of this 'transnationalism by proxy' is that it helps to build social capital. In the briefest possible way, the conventional wisdom regarding social capital can be summed up as follows: '*It's not what you know, it's who you know*' (Woolcock and Narayan 2000). The basic idea is that a person's social network constitutes a value in all walks of life, and that what is true for individuals also holds for groups. Communities or neighbourhoods with active networks and organizations are differently positioned when it comes to facing crises, solving problems, and taking advantage of new opportunities than those whose common arenas and social interaction are lacking (see also Chapter 2 in this volume). More social capital increases the capacity for collective action.

In Norway, the position of the LTTE has enhanced 'enforceable trust' as a social capital-generating mechanism. Important to this mechanism as defined by Portes and Sensenbrenner (1993, 1337) is the degree of internal 'reachability' of the members of the group. What is central is the possibility, and ability, to keep people in line and mobilize them for common goals. This requires both a common standard of behaviour and means of communication (Fuglerud and Engebrigtsen 2006).

My argument on social capital is in line with – but not exactly similar to – perspectives on immigration to the USA set forth in the works of Portes and Rumbaut (e.g. Portes 1996; Portes and Rumbaut 2001). The starting point for their research is that immigrants on arrival in the US today face a different society than the migrants of European origin did a hundred years earlier, and so, therefore, do their children. Some groups among today's second generation may well go through a smooth transition into the mainstream society and keep ethnicity as a matter of personal choice, while others need to muscle their way up, socially and economically, based on their own community's networks and resources. Still others must carry their ethnic identity as a mark of subordination and will be at risk for marginalization in America's inner cities.

Portes and Rumbaut argue (2001) that one important resource necessary to promote educational success and ward off the threats posed by discrimination, narrow labour market options, and street culture is social capital in the form of family structures and ethnic networks. The two share Coleman's (1988) view that one of the most important assets one can impart to children, not only emotionally but also in terms of social mobility, is growing up in a two-parent family. For the unit of the family, the outside environment in the form of the co-ethnic community determines the level of social capital available. This social capital is crucial in two respects: first, it tends to increase the economic opportunities available to immigrant parents, and, second, strong ethnic communities normally enforce norms against divorce and marital disruption, thus helping to preserve families. Particularly commendable in Portes and Rumbaut's work is the fact that it deals not only with the positive, but also with the negative effects of social capital. One of the challenges facing children of immigrants is a measure of solidarity among the marginalized, translating into a denigration of schools and teachers' instruments of racial oppression, and of education itself as incapable of bettering their situation (2001, 60). This oppositional ideology often ends up reinforcing the blockage of opportunities that it denounces.

In the case of Tamils in Norway, one may observe this ambiguous role of social border maintenance both among people who have migrated themselves and among their children. As for the second generation, the high level of social capital in the parental generation, stemming at least in part from the structuration of transnational loyalties, seems so far to have served their children well. One measure of this is the exceptionally low rate of divorce and, therefore, the high rate of Tamil children living in nuclear families (Øia 2007). This reluctance on the part of parents to separate is probably the result a Tamil 'value introjection' (Portes and Sensenbrenner 1993) regarding marriage. But it also seems reasonable to argue that it is in part a result of the organization of social relationships found in the Norwegian Tamil community. Based on the hypotheses put forward within social network theory (Kapferer 1973) it seems reasonable to suggest that the overlapping networks found among Tamil spouses, consisting of common friends and family of the two, makes it more difficult to separate. The importance Portes and Rumbaut afford this form of family life (mother, father, children) is confirmed by the fact that Tamil children are doing better than non-immigrant children in both elementary and higher education. This is further solidified by the fact that compared to both Norwegian and non-western immigrants, very few Tamil youths in the *Young in Oslo*-sample report (Øia 2007) had experiences with drugs.

Success with a cost

The resilience of the Tamil community – and of Tamil families – in Norway comes at a cost. In general Tamil youth seem to have a wish to become more deeply embedded in Norwegian society than what is practically possible, at least in part because of their parents and the level of social control within the Tamil community. This is especially the case for girls – as many as 95 per cent of girls and 82 per cent of boys of Tamil background in the *Young in Oslo* sample say that they experience themselves as Norwegian, which is significantly higher than the non-western immigrant average (77–9 per cent). While taking part in Tamil traditions, and having a sense of belonging to the ethnic community, a very significant part of Tamil youth, and the girls especially, say that they like to 'hang out' with people from other communities.

We do not have very good statistical measures for the relationship between social control and feelings of isolation from mainstream society. However, to a certain extent this link can be deducted by seeing statistics and interviews together. In terms of social control, what we know from the survey material is that 85 per cent of girls and 69 per cent of boys say that their parents know most of their friends and whom they spend their free time with. One of the most interesting aspects of these figures is the gender difference between girls and boys, something that also came out very clearly in the interviews. In general Tamil girls take it as a natural fact of life that their existence is restricted and confined in a different way than that of their brothers. Many reported not being allowed to go out or meet friends after school at all. And many expressed a feeling of being under surveillance not only by parents, but also by 'the big Tamil eye' – an actual expression used by one – including gossip, rumours, and slander. In a typical response to our observation that the community seems to be very tight, one Tamil girl says:

> Oh yeah, it's like every time you move you see the looks and the stares [laughing a little], blah, blah, blah. Yes, they call our home immediately and [say] 'your daughter is out with that one and that one, she's there and there'. Oh, please! [laughing]. It's because they are so afraid, you know. My parents have actually become quite liberal, but it is ... now I don't bother to go out so much, anyway, because now I'm afraid of the others, you know. It's not my parents that are the problem – I just can't stand listening to the others.

While many Tamil parents, perhaps, would see the reactions of this girl as the intended and successful result of a strict upbringing, there are at least two aspects of the situation that are worrying in the long term. One is that the parental enclave form of integration, with its reluctance to permit social networks outside the community, may limit the possibilities of the next generation with respect to social mobility in mainstream society. In other words, unless parents want their children to continue in the unskilled sectors where they themselves are employed, the conditions for successful participation in mainstream society may be different for their children than for themselves.

The other reason for concern is that for the youngsters, and in particular for the girls, the very strict upbringing and the difficulty of having a social life outside the family seems to lead to a sense of loneliness and depression. Looking at the *Young in Oslo* statistics, we see that 15 per cent of Tamil girls and 14 per cent of Tamil boys report 'often or very often feeling lonely', a number significantly higher than the non-western immigrant average. Further, only 44 per cent of Tamil girls report that the statement 'I feel that my peers accept me' is correct for them. That is significantly below all other groups. Finally, 22 per cent of Tamil girls report having tried suicide once or several times. These numbers point to a rigidity in community organization that represents a challenge to the generation now growing up in Norway.

Transcontinuity: The larger picture

I mentioned above that my perspective is like but not identical with Portes and Rumbaut's analysis of immigration to the USA (2001). Where I differ from their perspective is in seeing the social and cultural baggage that immigrants carry with them on arrival as more important. Portes and Rumbaut's focus is on the way the reception of immigrants, and shared immigrant experiences in this respect, structure future possibilities in the new country, also for engaging in transnational economic activities (Portes, 2001). However, I am concerned with what has been termed 'transcontinuity' which I see as connected to the hierarchical social imagery mentioned at the opening of this chapter. Schlee (2001) defines transcontinuity as the presence of '(e)lements of a social structure, or a political system, which survive revolutionary alterations and always re-emerge, albeit perhaps under a different name and guise, whatever social ruptures may occur' (18). More specifically I am interested in the social background of migrants and the 'possibility-structure' this brings in a longitudinal and trans-generational perspective (see also Chapter 4 in this volume). I agree with Schlee that continuities can and should be sought not only in larger social formations but also in the organization of extended families, in gender roles, relationships between parents and children, etc. One reason for this interest in transcontinuity is my wish to look at migration not only from a Norway- and Europe-centric perspective of 'integration' but from a perspective considering the long-term consequences of transnational processes to the migrating society as such. One way of approaching this issue is to look at the way in which migration strengthens or mitigates already existing social divisions in society. Transcontinuities are part of such processes.

While the picture is difficult to document in detail, there are many indications that migration from Sri Lanka to the west, also when the war was at its worst, has been structured by pre-existing caste and class differences. A brief introduction to pre-war Jaffna society is necessary to see how this works.

Being part of the Indian cultural region, society in Ceylon (Sri Lanka) has since time immemorial been organized around the principle of caste hierarchy. 'Caste' as principle of organization is based on the hereditary transmission of occupations and duties of service. These occupations and services are carriers of social status and rank. On the Sinhalese side, caste differences were primarily a form of secular ranking upheld by the state (Pieris 1956). Among Tamil-speaking people, the majority being Hindu, castes had and still have the religious and ritual dimensions known from the Indian mainland. The caste system found in Jaffna, the cultural centre of Eelam Tamils and the area where most international Tamil migrants come from, is the most rigid, with the most clearly defined patterns of inequality, discrimination, and social rejection among the varieties found in the island. According to Pfaffenberger (1982), 'Jaffna, so far from representing a marginal area of little comparative interest, is ... an extremely conservative and very South Indian region which to this day preserves ... the very heart of the South Indian cultural design' (28). However, there are also differences from the pattern common in South India. The social system in pre-war Jaffna was dominated by the landowning caste, the Vellalars, constituting approximately 50 per cent of the population. Brahmins, the caste in India normally considered the most 'pure' and therefore the highest ranked, have in Jaffna been relegated to providers of ritual services in temples owned by landowners.

The lower ranks of the caste hierarchy are in Jaffna referred to as 'Panchamar' – a collective label referring to the castes of Vannar (washers), Ampattar (barbers), Pallar (labourers), Navalar (toddy tappers),[3] and Parayar (funeral drummers). The latter four were, and to some extent still are, considered 'untouchable'. Traditionally the Panchamar castes constituted approximately 18 per cent of the Jaffna population. They were subjected to a range of restrictions and discriminations, ranging from prohibitions against respectable clothing to restrictions on the use of public space, access to drinking water, and entrance to temples and tea shops (Silva, Sivapragasam and Tanges 2009).

One restriction of interest to the present discussion was the prohibition of education (Silva, Sivapragasam and Tanges 2009, 57). The establishment

3 'Toddy tappers' are people who tap sap from the Palmyra palms to produce a fermented drink called 'toddy'.

of mission schools in Jaffna during the colonial period did not ameliorate the existing social hierarchy; rather the opposite. As noted by Silva, Sivapragasam and Tanges (2009):

> During the colonial period, whatever new opportunities opened up for advancement (such as education, business, commercial farming and government employment) were grabbed by the Vellalar families who were already privileged with a resulting intensification of caste disparities and a solidification of the caste system. (50–1)

Similarly, Mahroof (2000) observes that

> Education, and further education, thus, had a two-fold effect on the caste system of north Ceylon. In the case of the Vellala, it gave them even more earning power, status, and power to control themselves and those in the environment (which they by birth, had already in some measure). In the case of the deprived castes, it marginalized them still further. By interposing a new language, English, it created two worlds; the world of the rich Vellala and the depressed castes, whose only communication was in Tamil, unknown to most of the Ceylonese, including Sinhalese. (48)

Not that discrimination went uncontested. Panchamar protests against discrimination started in the 1920s and continued right into the era of the civil war. The first major issue became the 'equal seating' campaign, fighting for the right of pupils considered low caste to sit on benches along with high-caste children in classrooms instead of on the floor. After two years, administrative permission to do so was given, resulting in high-caste protesters burning down some fifteen schools that implemented the new regulation (Silva, Sivapragasam and Tanges 2009, 58). The last major issue, playing out in the 1960s, was the question of temple entry for 'untouchable' castes, culminating in a violent confrontation outside Maviddapuram Kandacami Temple in 1968 when dozens of self-styled 'defenders of Saivism', Hindus of high rank and their servants attacked peaceful Panchamars with iron rods and sand-filled bottles, the latter demonstrating for their right to enter the temple (Pfaffenberger 1990). While the Maviddapuram incident was the most violent, similar skirmishes continued into the late 1970s.

The relationship between caste conflicts and the later struggle for Tamil independence that has led to approximately half of the pre-war Tamil-speaking population now living outside Sri Lanka is a very interesting subject for discussion, for which there is no space here (see Pfaffenberger 1990; Hellman-Rajanayagam 1993; Fuglerud 2009, 2011). What should be noted, however, is that in the late 1960s many of the migrants now settled in Norway and other western countries were in the ten- to fifteen-year-old age range – in other words, the caste conflicts of that time were part of their own life-worlds and that of their parents. A few years later, many of the children of the 1950s and 1960s would be shipped out by their Vellalar middle-class parents to prevent them from being conscripted by the LTTE. My argument is, in short, that like the opportunities under the colonial regime mentioned by Silva, Sivapragasam and Tanges above, refugee-migration became if not an opportunity, then a solution grasped by the upper and middle castes. Siddhartan (2003) observes that

> the middle class also began to migrate to Colombo with the commencement of Eelam war III [1995]. It is this group, which is largely dependent on financial assistance from their children or other family members living abroad. Their main objective is to migrate to foreign countries, either legally or illegally, with the support of their relatives or family members living abroad. (310)

I would argue that the process Siddhartan describes commenced somewhat earlier, but I agree with the general picture. When collecting data for my book about Tamil refugees in Norway in the early 1990s, I concluded that 'members of Vellalar, Karayar, Ceviyar, and Thimilar castes probably make up more than 80 percent of the population in Norway' (Fuglerud 1999, 86). None of the non-Vellalar castes mentioned belong to the Panchamar category – they are 'middle-castes'. Similarly, Madavan in a study from 2011 of population movements in Jaffna notes that

> By linking this map with that of the traditional location of castes in Jaffna ... we could notice that it was especially areas traditionally occupied by the Vellalar which lost people more often, at least 500 people in three years. The Vellalar community began to leave the peninsula early, thanks to its financial resources, to get away to the south or abroad. (Madavan 2011, 14)

What we find in the Eelam Tamil situation is a 'division of labour' where the upper and middle layers of society have left the country to support the liberation struggle economically, while what are regarded as the lower layers remained to do the actual fighting (Fuglerud 2009). The Eelam Tamil case, therefore, illustrates very clearly the relationship between mobility and immobility pointed to by Glick Schiller and Salazar (2013). The division of labour was not based on consensual agreement; it was a distribution of tasks decided and enforced by the LTTE; coming into power they left no room in Jaffna for the traditional elite. As noted by Silva, Sivapragasam and Tanges (2009),

> while the LTTE eliminated more or less the traditional Vellalar political leadership in Jaffna through its campaign of terror, there were many waves of selective outmigration of 'high- caste' and 'high class' Vellalar families from Jaffna ... (60)

At the same time, the LTTE, through their own strict exit-visa regulations, actively prevented people considered 'fighting material' from leaving the territory under their control. This larger picture has also been noted by S. R. H. Hoole[4] in his remarkable ethnographic fictional story, 'The exile returned' (Hoole 1997):

> While those like Varnalingam and Rajaratnam in the ghetto [that is, the diaspora community] were praising the war for destroying caste, here [in Sri Lanka] the lower castes were worse off. Before, they simply labored under the Vellalahs; occasionally their women were ravished by the Vellahlahs. But now, they were dying for the Vellahlahs ... And through their death, the Vellahlahs in London, New York, Los Angeles, Toronto, Melbourne and Sydney, were receiving 'racial pride'. The Vellahlahs were in fact playing it both ways. Some were supporting the Leader [that is, Prabakaran, the leader of LTTE] in the west and, at the same time, going to their Sinhalese friends for help to establish businesses in Colombo for themselves. Other Vellahlahs in Ceylon were shuttling between Colombo and Jaffna, saying 'Sir' to Sinhalese ministers with whom they kept up their friendships, and, in Jaffna, asking for favours and giving the Leader economic advice and making fiery speeches, thereby

4 Hoole is a well-known academic and intellectual in Sri Lanka, intimately familiar with Tamil society in Sri Lanka and in the diaspora. In his academic career, he has alternated between posts in Sri Lanka, Singapore, and the USA. In 2006, Hoole was elected as one of three candidates for the post of Vice Chancellor of the University of Jaffna, and was subsequently appointed to this position by the president of Sri Lanka. Due to threats from the LTTE he was unable to take on this post and returned to the USA.

> protecting their houses in the North from being nationalized for the great cause. Whoever won they would be on top. (317)

This process of selective emigration is normally not considered when studying migration from the vantage point of settlement countries. It is, however, of great importance to the Tamil society at large and has over the many years of conflict positioned the diaspora and the homeland society vis-à-vis each other. To the extent that the Eelam Tamil society in the future remains *one* transnational social formation, with the economic and educational resources available in western countries, there is reason to predict that the traditional elite within this formation will regain and strengthen their dominance, particularly so with the decimation of the LTTE as an Eelam-based organization in 2009. To some extent we already see this happening with a shift in the centre of gravity in Tamil politics from the 'homeland' to the diaspora (Brun and Van Hear 2012). The fact that this increased influence of the diaspora is exerted in the name of keeping the political legacy of the LTTE alive, by those who fled their country to avoid having to fight for these ideas, is only one among many paradoxes of the Sri Lankan national question and does not change the course of this development.

Bonding, bridging and linking social capital

In this chapter I am not concerned with politics in a narrow sense, but with social transformation and development. In terms of social capital, Tamils in Norway have been able to forge a bonding horizontal solidarity among upper and middle castes, anchored in a concept of national identity that did not exist in the pre-war hierarchical society. Indications are that the same is happening in other settlement countries. It is important to remember that traditionally, castes, even upper castes, did not operate based on class interests; interests were organized vertically between separate land-owning families and their labourers and service-caste dependents. The refugee-migration of the last three decades has caused the development of a class-dimension to this hierarchical organization – a co-ordination of interests between members of the upper levels of society. In the local Norwegian context, this has involved the transformation of bridging networks between people of different families and villages into bonding social capital. With the qualifications pointed to above this has so far served them well as a group.

This in-group identification seems to be passed on to the next generation through families. What Engebrigtsen and I found in our interviews with youth of Tamil background, most of them in the fifteen- to twenty-two-year-old age span, was a strong sense of identification with and loyalty towards their parents. We also found that the Tamil middle-class valuation of education has been internalized by the young. Expressing a strong wish to be part of Norwegian society, youth of Tamil background in general seem to perceive their parents' migration to Norway, and their life situation there, as a sacrifice on behalf of the next generation. Moreover, the youth in general see this migration as a debt to be repaid to their parents by educating themselves and by behaving well per the standards of the Tamil community. Education is not seen by the youth primarily as personal qualification, or something to which they have an individual right as a citizen, but – even if not expressed in those exact words – as a collective endeavour, as part of a project of social mobility of the family and the Tamil community as a whole. For example, we asked a fourteen-year-old girl of Tamil background why she though school was important. She answered,

> You know, to have an education, not an advanced education but an education ... it helps you a lot in the future, doesn't it? When you're a foreigner, a Tamil, then education is pretty important. Normally Tamils, living in Asia, in those areas, they think that school is everything and I also have something from there, from parents who say that school is everything ... I also have to say that, don't I? Because it affects me when they say that if you lose control here there is no reason for you to be here. They tell me that – so, that's how it is.

'If you lose control, there is no reason for you to be here' – that is the message from parents and the Tamil community to the next generation. Another Tamil girl reflected,

> Well, the thing is, I think there is something special with us Tamils. Because, I figure, if it hadn't been for us children our parents wouldn't have been here. Because of this I feel that ... they live here, they have sacrificed so much, so we should also sacrifice something. Just to make them a bit happy ... So, I think we should be a bit more understanding and show them respect. And we should know what they are going through. They don't say 'yes, I miss my country' or 'I miss my mother'. But we should understand that without them saying it. That's what I think.

The people left behind

What about those who have not migrated? Due to heavy militarization, the possibility of doing research on the ground is limited in Sri Lanka's northern parts. However, there is no doubt that six years after the sudden decimation of the LTTE the general situation is difficult. IRIN, the news service of the UN Office for the Coordination of Humanitarian Affairs, estimated in 2012 that out of the approximately 580,000 inhabitants in Jaffna District, 55 per cent of households lived under the poverty line of less than $1 a day (IRIN 2012). In 2010, the same organization reported 40,000 female-headed households in the northern region, approximately 20,000 of them in Jaffna District, with 50 per cent of the females heading the families being under thirty years of age (IRIN 2010). This is in stark contrast to the families living in Norway. BBC Asia reported on 30 March 2014 that there are 31,524 people still living in IDP camps around Jaffna because their land is held by the military. In addition, there are large numbers of internally displaced persons living with host families (IRIN 2012). As an indication of how many, the ratio between IDPs living with relatives and friends to IDPs living in camps in Jaffna District in 2009 was ten to one (Silva, Sivapragasam and Tanges 2009, 65).

While accurate information is not available about the caste background of IDPs in Jaffna, or of the remaining population in general, indications support the hypothesis that a majority of those remaining in camps are from castes considered 'lower'. For example, in a study conducted by Silva, Sivapragasam and Tanges in Mallakam in 2007, which is on the northern coast of Jaffna, the totality of people languishing in welfare centres were of Panchamar origin, although these castes comprised only 58 per cent of the pre-war population (Silva, Sivapragasam and Tanges 2009, 68). Among the reasons for this situation pointed to by the authors, is the lack of social networks outside their own communities on the part of depressed castes, something that restricts their capacity to move out of the camps and also to acquire land. Traditionally the lower castes did not own much land themselves, but were working on land owned by Vellalars. Today, even if they have the resources, they are prevented from buying land by caste discrimination.

In terms of social capital, I suggest that what has happened in Jaffna in the course of the war is that lower castes have experienced a drastic reduction in their linking capital. The emigration of traditional landowners and the decimation of the de facto state of the LTTE have both contributed to this. The notion of 'linking capital' is a conceptual refinement of social capital theory defined by Szreter and Woolcock (2004) as 'norms of respect, and networks of trusting relationships between people who are interacting across explicit, formal or institutionalized power, or authority gradients in society' (655). Whereas established concepts of bonding and bridging social capital are primarily horizontal metaphors for closed versus open social groups, linking capital draws attention to vertical power asymmetries and, in non-technical jargon, the importance of 'having friends in high places'. To what extent the current situation may in time lead to the generation of new forms of social organization – horizontal or vertical – remains to be seen. At the moment, the prospects are bleak.

Conclusion

The relationship between host country and country of origin orientation is complex issue and one should be careful not to draw hasty conclusions. It is also a relationship that may change over time, dependent on state policies, media profiles, and the social mobility of the people in question (cf. Glick Schiller and Salazar 2013, Engebrigtsen in this volume). These factors feed into ideological constructions of refugee groups through the public image and through schemes of surveillance, control, and administration (cf. Ong 1996). In the case of Tamil refugee-migrants to Norway their labelling as 'terrorists' in the early phase was used for physical and social exclusion. This situation seems, at first glance, to fit well with a diasporic and transnational perspective. In public discourse and to some extent in academic literature, there has been a tendency to regard the two processes of integration and transnational engagement – one internal the other external – as forces pulling in opposite directions. This perception is clear in one of the early works within this tradition, Safran's (1991) definitional essay on diasporas, where one central criteria – Safran's third – is that members of a diaspora 'believe that they are not – and perhaps cannot be – fully accepted by their host society and therefore feel partly alienated and insulated from it' (83–4). Anderson's essay on long-distance nationalism from the same period (Anderson 1992) makes a similar, if more general, argument when pointing out that while refugees in the nineteenth century often took active part in the nation-building of societies where they settled, today there is a tendency for refugees to isolate themselves from their new social surroundings with the country of origin politics remaining the focus of their attention. Incidentally, I have myself made the same point in an earlier work on Tamil refugees in Norway (Fuglerud 1999), arguing that the alienation ensuing from a strict immigration regime strengthened diasporic identifications. While I do not feel any need to go back on this assessment, the years that have passed have added complexity to the situation. In fact, I now argue that the homeland orientation has over time turned out to help *overcome* exclusion and has promoted social mobility.

Regarding migrants' position vis-à-vis their societies of origin I fully agree with Glick Schillller and Salazar (2012) that too little attention has been paid both to the relationship between mobility and immobility and to the ways in which this relationship may articulate class formations. In part this is because the analytics of 'diaspora' as deployed in the social sciences often retain a certain host-state bias. Mainly concerned with the reproduction of solidarity across borders the use of the concept of diaspora for analytical purposes also often rest uneasily on assumptions of homogenous and static political collectives based on ethno-nations, paying too little attention to internal divisions (cf. Anthias 1998). Reading the relevant literature one may be led to think that when a transnational social field is established, spanning homeland and settlement countries, this will benefit economic development for all and encourage a sense of common identity between members of the national community. In the case of Eelam Tamils, I regard this scenario as uncertain at best.

Bibliography

Anderson, Benedict. 1992. 'Long-distance Nationalism. World Capitalism and the Rise of Identity Politics'. The Wertheim Lecture 1992. Amsterdam: Center for Asian Studies Amsterdam.

Anthias, Floya. 1998. 'Evaluating Diaspora: Beyond ethnicity?' *Sociology*. 32 (4): 557–80.

Bakewell, Oliver. 2008. 'Research Beyond the Categories: The Importance of Policy Irrelevant Research into Forced Migration'. *Journal of Refugee Studies*, 21 (4): 432–53.

Black, Richard. 2001. 'Fifty Years of Refugee Studies: From Theory to Policy'. *International Migration Review*, 35 (1): 57–78.

Brubaker, Rogers S. 2005. 'The "Diaspora" Diaspora'. *Ethnic and Racial Studies*, 28 (1): 1–19.

Brun, Catherine, and Nicholas Van Hear. 2012. 'Between the Local and the Diasporic: The Shifting Centre of Gravity in War-torn Sri Lanka's Transnational Politics'. *Contemporary South Asia*, 20 (1): 61–75.

Cels, Johan. 1989. 'Responses of European States to de facto Refugees'. In *Refugees and International Relations*, edited by Gil Loescher and Laila Monahan, 187–215. Oxford: Oxford University Press.

Coleman, James. 1988. 'Social Capital in the Creation of Human Capital'. *American Journal of Sociology*, 94, Supplement: Organizations and Institutions: Sociological and Economic Approaches to the Analysis of Social Structure: 95–120.

Engebrigtsen, Ada, and Øivind Fuglerud. 2007. *Ungdom i flyktningfamilier. Familie og Vennskap – Trygghet Eller Frihet?* [*Youth in Refugee Families. Family and Friendship – Safety or Freedom?*]. NOVA Rapport, 7 March.

——2009. *Kultur og Generasjon. Tilpasningsprosesser Blant Somaliere og Tamiler i Norge*. Oslo: Universitetsforlaget.

Faist, Thomas. 2013. 'The Mobility Turn: A New Paradigm for the Social Sciences?' *Ethnic and Racial Studies*, 36 (11): 1637–46.

Fuglerud, Øivind. 1994. '"Working for my Sisters ...": Tamil Life on the 71st Parallel'. *Migration*, 23 (4): 87–103

——1997. 'Ambivalent Incorporation: Norwegian Policy Towards Tamil Asylum-seekers from Sri Lanka'. *Journal of Refugee Studies*, 10 (4): 443–61

——1999. *Life on the Outside. The Tamil Diaspora and Long Distance Nationalism*. London: Pluto Press.

——2001. 'Time and Space in the Sri Lanka-Tamil Diaspora'. *Nations and Nationalism*, 7 (2): 195–215.

——2002. 'The Housewife and the Soldier. Tamil Constructions of Womanhood in the age of Migration'. *Kvinder, Køn & Forskning*, 2: 20–30.

——2003. 'Local Communities and State Ambitions in Eastern Sri Lanka'. In *Building Local Capacities for Peace: Rethinking Conflict and Development in Sri Lanka*, edited by Markus Mayer, Darini Rajasingham-Senanayake and Yuvi Tangarajah, 65–79. New Delhi: Macmillan India.

——2004a. 'Space and Movement in the Sri Lankan Conflict'. In *Refugees and the transformation of societies: Loss and recovery*, edited by Philomena Essed, Georg Frerks and Joke Schrijvers, 42–52. London: Berghahn Books.

——2004b. 'Constructing Exclusion'. *Social Anthropology*, 12 (1): 1–16.

——2009. 'Fractured Sovereignty: LTTE's State-building in a Global World'. In *Spatialising politics: Culture, difference and geography in postcolonial Sri Lanka*, edited by C. Brun and T. Jazeel, 194–215. India: Sage India.

——2011. 'Aesthetics of Martyrdom: The Celebration of Violent Death Among the Liberation Tigers of Tamil Eelam'. In *Violence Expressed: An Anthropological Approach*, edited by Nerina Weiss and Maria Six-Hohenbalken, 71–90. Farnham: Ashgate Publishing.

Fuglerud, Øivind and Ada Engebrigtsen. 2006. 'Culture, Networks and Social Capital: Tamil and Somali Immigrants in Norway'. *Ethnic and Racial Studies*, 29 (6): 1118–34

Glick Schiller, Nina, and N. B. Salazar. 2013. 'Regimes of Mobility Across the Globe'. *Journal of Ethnic and Migration Studies*, 39 (2): 183–200.

Glick Schiller, Nina, Linda Basch, and Cristina Szanton Blanc. 1992. *Towards a Transnational Perspective on Migration: Race, Class, Ethnicity, and Nationalism Reconsidered*. New York: Annals of the New York Academy of Sciences.

Hellman-Rajanayagam, Dagmar. 1993. 'Jaffna Social System: Continuity and Change Under Conditions of war'. *Internationales Asienforum*, 24 (3–4): 251–81.

Hoole, S. Ratnajeevan H. 1997. *The Exile Returned: A Self-portrait of the Tamil Vellahlahs of Jaffna, Sri Lanka*. Colombo: Aruvi Publishers.

IRIN, 2010. 'Sri Lanka: Women Take Over as Breadwinners in North'. <http://www.irinnews.org/report/90429/sri-lanka-women-take-over-as-breadwinners-in-north> accessed 20 May.

——2012. 'Sri Lanka: Focus on Food Insecurity in Jaffna'. <http://www.irinnews.org/report/95828/sri-lanka-focus-on-food-insecurity-in-jaffna> accessed 15 June 2015.

Kapferer, Bruce. 1973. 'Social Networks and Conjugal Roles in Urban Zambia: Towards a Reformulation of the Bott Hypothesis'. In *Network Analysis: Studies in Human Interaction*, edited by Jeremy A. Boissevain and James Clyde Mitchell, 83–110. Paris: Mouton.

Landau, Loren B. 2007. 'Can we Talk and is Anybody Listening? Reflections on IASFM 10, "Talking Across Borders: New Dialogues in Forced Migration"'. *Journal of Refugee Studies*, 20 (3): 335–48.

Levitt, Peggy, and Nina Glick Schiller. 2004. 'Conceptualizing Simultaneity: A Transnational Social Field Perspective on Society'. *International Migration Review*, 38 (3): 1002–39.

Madavan, Delon. 2011. 'Socio-religious Desegregation in an Immediate Postwar Town Jaffna, Sri Lanka'. *Carnets de Géographes*, no. 2: 1–22.

Mahroof, M. M. M. 2000. 'A Conspectus of Tamil Caste Systems in Sri Lanka: Away From a Parataxis'. *Social Scientist*, 28 (11/12): 40–59.

Øia, Tormod. 2007. *Ung i Oslo – Levekår og Sosiale Forskjeller* [*Young in Oslo – Living Conditions and Social Differences*]. Report 6/2007. Oslo: Norwegian Social Research.

Ong, Aihwa. 1996. 'Cultural Citizenship as Subject-Making'. *Current Anthropology*, 37 (5): 737–61.

Pfaffenberger, Bryan. 1982. *Caste in Tamil Culture. The Religious Foundations of Sudra Domination in Tamil Sri Lanka*. Syracuse: Syracuse University.

——1990. 'The Political Construction of Defensive Nationalism: The 1968 Temple-Entry Crisis in Northern Sri Lanka'. *The Journal of Asian Studies*, 49 (1): 78–96.

Pieris, Ralph. 1956. *Sinhalese Social Organization: The Kandyan Period*. Colombo: University of Ceylon Press.

Portes, Alejandro (ed.). 1996. *The new Second Generation*. New York: Russel Sage Foundation.

Portes, Alejandro. 2001. 'Introduction: The Debates and Significance of Immigrant Transnationalism'. *Global Networks*, 1 (3): 181–93.

Portes, Alejandro, and J. Sensenbrenner. 1993. 'Embeddedness and Immigration: Notes on the Social Determinants of Economic Action'. *American Journal of Sociology*, 98 (6): 1320–50.

Portes, Alejandro, and Ruben Rumbaut. 2001. *Legacies. The Story of the Immigrant Second Generation.* Berkeley: University of California Press.

Safran, William, 1991. 'Diasporas in Modern Societies: Myths of Homeland and Return'. *Diaspora*, 1 (1): 83–101.

Schlee, Günther. 2001. 'Regularity in Chaos: The Politics of Difference in the Recent History of Somalia'. Working paper no. 18, Max Planck Institute for Social Anthropology: Halle.

Siddhartan, M. 2003. 'Negotiating "Tamilness": A Case Study of Jaffna Tamil Migrants to Colombo Since 1990'. In *Building Local Capacities for Peace. Rethinking Conflict and development in Sri Lanka*, edited by Markus Mayer, Darini Rajasingham-Senanayake and Yuvi Thangarajah, 305–20. Delhi: Macmillan India.

Silva, Kalinga Tudor, P. P. Sivapragasam and Paramsothy Tanges. 2009. *Casteless or Caste- blind? Dynamics of Concealed Caste Discrimination, Social Exclusion and Protest in Sri Lanka.* Colombo: International Dalit Solidarity Network.

Szreter, Simon and Michael Woolcock. 2004. 'Health by Association? Social Capital, Social Theory, and the Political Economy of Public Health'. *International Journal of Epidemiology*, 33 (4): 650–67.

SSB. 2007. *Fakta om 18 Innvandrergrupper i Norge.* [*Facts about 18 immigrant groups in* Norway]. Report 27/2009, Statistics Norway.

——2008. *Levekår blant innvandrere, 2005–6.* [*Living conditions of immigrants*]. Statistics Norway.

Taylor, Charles. 2002. 'Modern Social Imaginaries'. *Public Culture*, 14 (1): 91–124.

Turton, David. 2005. 'The Meaning of Place in a World of Movement: Lessons from Long-Term Field Research in Southern Ethiopia'. *Journal of Refugee Studies*, 18 (3): 258–80.

Wimmer, Andreas, and Nina Glick Schiller. 2002. 'Methodological Nationalism, the Social Sciences, and the Study of Migration: An Essay in Historical Epistemology'. *International Migration Review*, 37 (3): 576–610.

Woolcock, Michael, and Deepa Narayan. 2000. 'Social Capital: Implications for Development Theory, Research, and Policy'. *The World Bank Research Observer*, 15 (2): 225–49.

ADA I. ENGEBRIGTSEN

2 Mobile Subjects: Power Relations and Tactics for Survival

In a parking lot outside the Botanical garden in the centre of Oslo, twenty vans are parked side by side. On the fences around the garden, clothes are spread out to dry and while some women are busy frying chicken on small portable barbecues, others are washing dishes and others again are sitting around the fire eating. In some vans, one can see family members sleeping with the doors wide open, while some men are repairing cars or chatting and laughing in small groups. This scene resembles scenes in most Gypsy settlements I have seen in Eastern Europe. In Norway, where domestic privacy is strongly valued, this public display of private life is highly disturbing, challenging conceptual boundaries between self and other, public and private.

Romanies (Roma and other Gypsy populations) are present in most parts of the world. In Europe they make up large minorities in most countries. All European countries have both sedentary populations of Romanies that have been citizens for centuries and more recently settled populations of Roma. In addition, Romanian Romanies[1] that migrate as EU citizens and tourists and have since about mid-2000 travelled regularly between Romania and the rest of Europa as workers, beggars, and entertainers. Street begging and entertainment (music) are the most prominent forms of livelihood in Norway for migrating Romanies. Romanian Romany migrants to Norway and the semi-nomadic Norwegian Rom population are my cases for this discussion.

By looking at these two different Romany groups' struggle for survival historically and in the present, I will explore how the concepts of mobility

1 Roma and other similar minorities.

and nomadism can illuminate the power ratios between mobile populations and state formations. I will discuss how governments try to control 'mobile subjects' such as Romanies, and their responses hereto. I will discuss mobility in terms of the interrelatedness of physical movement and motivation for movement. I argue that in spite of their stigma as 'eternal strangers', Roma and Romanies are dependent on social connections both outside and inside their own group and thus have changing relationships both with Roma and non-Roma. The relationship between time and mobility and the relationship between sedentarism and nomadism is intrinsic for the discussion here (see also Chapter 4 in this volume).

Mobility, nomadism, mobile subjects

I employ the concepts of mobility and mobile subjects here as metaphors of the same order as the French political philosophers Deleuze and Guattari's implied in their work *Nomadology* (1986). Nomadology is a metaphorical concept and perspective for the aspects of the social world (groups, subjects, processes) that avoid state power, most often by being more or less mobile in a physical and/or psychological and intellectual sense. In nomadology the state represents the sedentary power with control over territory while the mobile subjects represent the de-territorialized subjects in opposition. In Deleuze and Guattari's political philosophy (1986, 1987), state power and nomadic power represent interdependent forces that are tied together through opposition. The nomad is metaphorically termed *a war machine against the state* because she defies classification and hence conservation and control. In this chapter, mobility and mobile subjects denote parallel social positions in a specific formation of power relations. While Deleuze and Guattari's nomads are foremost 'mentally mobile', mobility in this chapter implies the interconnection of physical and mental mobility. Mobility here is the position of a subject or group without a 'proper locus (place)' (de Certeau 1984, 37) who resides on the territory of others, with all that denotes about the ability to plan ahead, control time and to take well-calculated actions. Mobile subjects are people who avoid being governed and who try to evade what they see as control that threatens their way of life and their communities. For the subjects and groups I am discussing, mobility is life-shaping (Salazar and Smart 2012). The combination of physical mobility and a mobile or nomadic habitus (mental mobility) is at the core of my discussion; an interconnection also emphasized by Hays and Ninkova in this volume. Whether an individual or a group/people can be understood to be 'mobile' in this power-sense is thus an empirical question. In this chapter, I claim this to be the position of the two Roma/Romany groups I discuss; they are mobile subjects – mobile by intentionally evading state control, and they are dependent on physical mobility for upholding their communities and their evasion of state power. Other mobile subjects and groups in this sense could be illegal migrants, religious dissidents, street-children, or international corporations outside state control (Deleuze and Guattari 1986; Urry 2014).

Migrating Romanies and Norwegian Roma

From the government's point of view, Romanian Romanies and Norwegian Roma in Norway resisting integration are culturally conservative groups, dominated by powerful males, and due to poverty or lack of education 'difficult to integrate'. The general idea is that they will change (or be salvaged) depending on the quality of the supportive projects developed for and bestowed onto them, expressions of what Foucault (as cited in Golder 2007) has called pastoral power. In Norway, soft or pastoral power is combined with more cohesive measures towards such 'nonconformists'. In the case of Norwegian Roma, cohesive measures are implemented in order to ensure children's school attendance and to secure proper socialization practices in Rom families. For Romanian Romanies, cohesive measures such as banning begging and 'rough sleeping'[2] and controlling mobility together with harsh police treatment (Nasjonal Institusjon for Menneskerettigheter

2 Sleeping outdoors in the city.

2015) are applied. In both cases, the government sees physical mobility and illiteracy as the source of problems, and sedentarization and education as the solution. The Norwegian Roma must stop moving to ensure their children's education and adults' inclusion into the work force, and Romanian Romanies must go back to Romania and stay there in order to become integrated Roma/Romany citizens and avoid poverty. However, this solution rests on a misinterpretation of the life-world, social position, and economy of most Roma and Romany populations.

Tactics and strategies

Inspired by de Certeau's (1984) discussion of strategies and tactics in power struggles, the analytical idea underlying this chapter is that mobile subjects and networks have instruments at their disposal for exerting power other than territorialized subjects and institutions. In de Certeau's analytical model, tactics are 'weapons of the weak', or, more precisely, weapons of the peoples without territory, while strategies are only available to subjects and groups with territorial rights. I further claim that the concept of resistance as an opposition to domination does not capture the subtle ongoing process of negotiation of dependency and autonomy. I do not see the Roma's acts of avoidance and ambiguity as resistance and opposition to domination in Scott's (1990) sense, but as outcomes of the relationship between two modes of existence and of power; the 'nomad mode' and the 'state mode' (Deleuze and Guattari 1986), and as expressions of their specific strategic position and of the Rom habitus. Consequently, I argue that the Roma's troubled relationship with authorities is an unforeseen outcome of their past experiences and of the way they form communities based on specific modes of subsistence, organization, power structure, and cosmology (Engebrigtsen 2007). To understand the particular situation of these highly diverse groups of Roma/Romanies, my main question is: what is the strategic situation of most Roma groups, what weapons do they have to their disposal, and what are their options for securing relative cultural autonomy? Before developing this perspective further, I will present the data for this study and the historical background of the cultural traits of the two Roma populations.

Methods

Data for this chapter is based on my PhD work on Roma and peasant relations in a Romanian village in 1996/7 and on my two studies regarding migrating Roma and other similar groups as beggars and street-workers in Norway in 2009–12 and 2013–14. My data of Roma in Romania are based on a one-year participant observation – participation in daily life as well as in ritual manifestations such as weddings and funerals together with informal conversations with men, women and children. The focus of my PhD study was the relationship between the Rom population and the non-Rom community (Engebrigtsen 2007). My data of migrating Romanian Roma and other Romanian citizens begging in Oslo is mostly based on interviews, observations, and informal discussions concerning their situation in Romania, their opportunities there and their experiences as beggars in Norway (Engebrigtsen 2012, Engebrigtsen, Fraenkel and Popet 2014). Data about the Norwegian Rom group are not based on research, but on seven years of participant observation as a kindergarten teacher in an open[3] Rom kindergarten in Oslo, and on reports and evaluations of the programme that the kindergarten was part of. These data are obviously of different quality, as they were produced in different contexts and times. They also concern different Roma populations at different times and in different places. Together they make up a comparative and multi-sited knowledge base. Interviews, conversations and observations with Roma in Romania and Norway were conducted in Romanian and Romanés, as my late husband spoke Romanés and we both spoke Romanian.

3 The kindergarten was open to parents and family members throughout the day.

The Roma … Who are they? Historical background

The Vlach Romanés-speaking Roma (Norwegian Roma and the majority of the Romanian street-working migrants in Norway) are the largest group of the highly diverse populations that today are generalized as 'Roma or Romanies' for political purposes. The term 'Vlach' is a linguistic term for the Romanés dialect spoken by the populations of Roma who entered Wallachia (the western and southern parts of present-day Romania) around the thirteenth century and Transylvania somewhat later. They were incorporated into the economy as slaves in Wallachia (Ţara Romaneasca) and serfs in Transylvania (then Hungary and the Austro-Hungarian Empire) (Achim 1998, 282).[4] All Roma and other 'Romany' groups were slaves in Wallachia and only these populations were slaves. They performed most of the forced manual labour in agriculture, in housework, as miners, and as artisans for their owners, and their owners were the church, the monasteries, royalty, and the nobility. Although these slaves were not legal subjects – in contrast to the almost entirely Romanian-speaking serfs, some Roma slaves, mostly the clans of smiths, could achieve relative independence and could make fortunes as long as they served their masters as well. Smiths and most of the other artisan slave groups were mobile in the physical sense and they travelled through their master's territory serving the serfs and masters with crafted tools (Achim). In Transylvania, Roma and other Romany groups where 'royal serfs' mostly mobile craftsmen, directly dependent on the royalty (43). The status of the Roma and other Romanies was quite distinct from other groups in Wallachia and Transylvania, but they lived in close interrelation with the Romanian and Hungarian speaking feudal peasants (44). After the abolition of serfdom and slavery around 1860, out-migration was legal and large parts of these populations migrated first to Western Europe and then to the rest of the world, were the majority became more or less sedentary. However, many stayed in Romania and were subject

4 Probably originating from the Indian territories some hundred years earlier, Romanés being classified as an Indic language.

to forced sedentarization and relative immobility in villages and towns, eventually under the Ceauşescu regime. Not until Romania was included as full member of the EU in 2006 could Romanians again travel out of the country, and this was the starting point of a second large emigration of Roma and other Romanies to Western Europe.

The Roma's history is a one of physical mobility, but also of people being constructed as strangers (Bauman 1991) residing on the territory of the Other. It is also a story of shifting regimes of mobility/sedentarism in different times (Glick Schiller and Salazar 2013). During times with forced sedentarization, many Romany groups became permanently sedentary, but others only shifted their patterns of mobility to adapt to political restrictions and took up mobile livelihood strategies as soon as new regimes made it possible. These are indispensable factors for understanding the cultural norms and ways of life of present-day Roma and other similar groups with background from Romania.

However, Romanies and Romanés speaking Roma are highly diverse when it comes to religion, way of life, and their inclusion in the majority society. Descendants of the Romanian, Romanés-speaking Roma who left Romania in the nineteenth century live in most countries of the world, often forming separate societies within the majority society, but always connected to majority society in different ways. But even other non-Rom populations of Romanies such as Sinti, Manush, Gitanos, Romanichel, or Travellers have long histories of persecution, exclusion, and discrimination in the countries where they have lived, histories that are crucial to their present political, economic, social, and cultural life.

Physical mobility is a viable way of life when resources are spread and marginal, and a way to evade political control. The political unpredictability that these populations have experienced creates time and space as an aspect of life to be adapted to rather than to control: moving means passing over the territories of others by adapting to the conditions that are set by them. Thus, the time/space relationship is different among mobile groups than among sedentary groups; mobile groups must relate to the moment, and act in the spur of the moment (de Certeau 1984). In a different sociocultural context, this is also a point made by Simonsen (see Chapter 6 in this volume). Most Roma groups in Romania are sedentary in the sense that they live in permanent villages, but the majority are still dependent on mobile strategies for their livelihood and for sustaining their communities. In contrast to farmers, their economic resources are spread over a wide territory, and may expand and diminish according to shifting power regimes. Romanies are in the same way sedentary, but they have historically been dependent on different modes of internal and external mobility for their economic survival and cultural and political independence.

Romanian Roma: Looking for greener pastures

My first case concerns Romanian Romany migrants in Norway, mainly village Roma and similar groups, and their relations with Norwegian authorities. Street-working Romanian migrants are almost entirely from the segments of the population called by the derogatory term *ţigani* in Romania, which is parallel to, but not synonymous with, the English word 'Gypsy'. They represent a range of different groups with similar histories and positions in Romania, with different self-ascription and different mother tongues. In the Second World War, the Roma and other Romanies who stayed in Romania risked forced deportation[5] and persecution. After the war and under the socialist government, outmigration was again prohibited, until Romania's inclusion in the EU in 2006. The largest group of the people termed *ţigani* in Romania today are self-defined Vlach Romanés-speaking Roma and descendants of those who stayed in Romania after the abolishment of serfdom and slavery. Others are self-defined Romanian-speaking *Roma Romanisat* or *ţigani* (Romanized Roma), define themselves by other group names, and some say they are just Romanian. The ethnic boundaries of Romanies in Romania are not fixed, but may vary according to time, place, and context. The Romanies who migrate today originate from different regions of Romania with a concentration around the central

5 Thousands of Romanian Romanies were deported to Transnistria, where thousands died of hunger and disease.

and eastern parts (Engebrigtsen 2012, Engebrigtsen, Fraenkel, and Popet 2014). In Romania, they often live on the fringes of villages or in a separate part of a village or town, but also in mixed villages with Romanians or Hungarians.[6] They engage in different activities from wage-labour, crafts, trade, farm work, gathering, and begging. Until the 1950s, several of these groups travelled in Romania with horse-pulled wagons in spring, summer, and autumn, gathering, exchanging, and performing simple services for the peasants. In winter they dug caves (*colibi*) outside the village to live, and stayed there until spring when they took up travelling again. Under Ceauşescu nomadism was banned, primary school was made mandatory for all Romanian children, and all men and most women were employed in state enterprises. After the revolution, Romany children were no longer expected to go to school, and families that wished to send their children were advised to send them to schools for children with special needs. Most Romanies lost their jobs and took up their old way of life, settled but mobile. As the resources they exploit are scarce, and found over large areas, mobility is a requirement in order to make ends meet. In some villages the women go train scavenging in a nearby town several days a week. The waste is fed to pigs that eventually are sold at the village market. Many work as farmhands for the peasants and several of the older women perform magic services such as palmistry and exorcism. The men may seek work or sell simple handmade products in nearby villages, or go on business ventures to other areas. Many are employed as day labourer in construction and roadwork. Families with horses may cut firewood for sale in the villages in the area (Engebrigtsen 2007).

It is the poor and unemployed segments of the Romany populations that go abroad 'looking for greener pastures' as it is termed in Romanés (Engebrigtsen 2007).[7] To most Romanian Romanies, going abroad simply means expanding one's mobile activities across national borders. They stay in Norway for a period before going back to Romania for some months

6 That is how they generally refer to themselves.

7 Romania also has a growing population of educated professional Romanies often working as lawyers in NGOs, etc., often for the benefit of Romanies, and a population of traditional, but wealthy Roma, who also migrate, but not to beg.

and then leave for Norway again. They travel in family/kin groups from the same village or community, mostly leaving their dependent children at home with relatives. This is a kind of seasonal migration that follows the weather, and the opportunities; few endure street-activities in mid-winter, but to those who do it is very profitable. The majority migrate in family and kin groups from the same village and town. They travel by rented minibuses, or by buss directly. As with all travel, several actors are eager to make money of Romany migration from Romania to the rest of Europe.[8]

As few young people have education or references from employment, they are almost entirely dependent on street-activities in Norway (Andriaenssens and Hendrickx 2011). They sit on pavements expecting alms, play music, sell magazines, or party stash to night owls in the city centre, gather empty bottles for recycling, some are employed in illegal economic activities, other in petty crime, and some engage in all these activities at different times. Romanian beggars in Norway practise 'rough sleeping' – some rent a mattress in overcrowded apartments, some sleep in cars, others in illegally occupied, run-down houses and under bridges, many sleep in tents in the woods surrounding the towns. The last years several shelters have been provided for by NGOs, were the homeless may sleep at a low cost. In Norway most homeless people are now Romanian. They do their laundry in the rivers and dry their clothes wherever possible, often on bridges or fences, they use dark corners, bushes, and back yards as bathrooms; in short, they display their daily life for all to see and thus challenge the boundary between the private and the public. Migrant Roma thus live their private life in public spaces; they occupy that space and make it theirs.

The authorities relate to what they see as 'trespassing' in different and often contradictory ways. The police and private security guards impose control and eviction. As it is illegal to sleep in public places in the city, the police and private security companies chase sleeping families, often several times in one night, and confiscate their belongings. Beggars who try to warm up in public buildings, like railway stations and shopping malls, are regularly

8 Romanies emigrate from all countries in Europe with large Romany populations, but in Norway almost all are from Romania.

chased and fined. A recently implemented regulation banns sleeping in cars in the city, a rule that primarily targets Romanian migrant beggars. However, this political and governmental aggression is paired with several other initiatives that soften their daily life: A street magazine is published for them, which enables them to make money without begging, even if security guards control where it can be sold. The shelters and outreach services, run by NGOs in the large towns, also represents a 'softening' of their harsh conditions.

There is an ongoing discussion among politicians about banning foreign beggars altogether, but there is strong opposition to this strategy by some politicians and parts of the public. Romanian beggars have supporters all over the country. Members of the public are concerned for their well-being and involve themselves through friendship, in housing projects, general economic support, wage-labour and other projects. The situation is rather ambiguous, the authorities try to control, Roma evade this control and challenge various classification such as between private and public and between work and charity.

Norwegian Roma: A national minority, surviving as Roma

The ancestors of the Norwegian Roma were most probably serfs or slaves in Romania/Hungary and started to travel in Norway (and to the rest of the world), after the abolishment of serfdom and slavery in the late 1800s. Like in Romania, they were dependent on spatial mobility for their livelihood. Many travelled all over Europe, but some stayed for long periods in Norway, and as church records show, married, had children, and were buried there. Their livelihoods were probably similar to present-days Romanian Roma migrants' – they were mobile crafts-men, peddlers and beggars. In the nineteenth century, Norway already had a relatively large population of travellers, non-Roma Romanies (tater, reisende, romani) who had settled in the seventeenth century as itinerant craftsmen, seasonal labourers and beggars (Hvinden 2000). In the late 1800s, a religious organization was commissioned by the state to settle and assimilate this population. The organization used harsh means to enforce settlement, wage-labour, and education (Halvorsen 2000; Hvinden 2000; Pettersen 2005). In the 1920s, when the organization started to show interest in the migrating Roma, they left the country (Engebrigtsen and Lidén 2010). When a group of Roma, with Norwegian passports, tried to re-enter Norway in 1934 in response to the tightening persecution in Germany, they were rejected and sent back to Germany/Belgium, from where they were deported to concentration camps where many of them perished (Johansen 1989; Rosvoll 2013). This was done with reference to 'Gypsy law' (sigøynerparagrafen), already launched in 1907, which prohibited Gypsies (sigøynere) from entering the country. After the war, around fifty survivors from this group with their families claimed asylum and were eventually, after several lawsuits, granted Norwegian citizenship (Hanisch 1976). In the 1950s and 1960s, they were living in caravans in a parking lot in the centre of Oslo surrounded by high fences. Human-rights activists contacted the group and, together with one family leader who titled himself a Gypsy king, they wrote a letter to the Norwegian king asking for help with integration into Norwegian society. Backed by government authorities, the municipality of Oslo planned a programme – the 'Rehabilitation of Norwegian Gypsies' – that aimed at settling the Roma in permanent houses, offering appropriate education for children and adults, vocational training for men, and a kindergarten for the children to enable the women to go to school. These activities were supported by social security funds; all families received allowances, adults received small wages for school attendance, and the kindergarten was free of charge (see also Chapter 4 in this volume). The programme was developed in co-operation with the Gypsy king – the influential but self-appointed Rom leader – and several other leaders of extended families. The programme represented a classical pastoral project in the Foucaultian sense (Foucault in Goulder 2007), aimed at changing self-understanding and mentality of Gypsies by voluntary sedentarization and education. After about ten years, most families were settled in houses that had been constructed for extended families in different townships in Oslo. The state still offered literacy education to all adults, paid by the hour. Some work-training projects were established and the municipality even started a carpet shop run by one family. Certain schools set up classes for Rom children and developed school material in *Romanés*, and some Roma were appointed as mother-tongue teachers. The municipality started a kindergarten for Rom children and a separate social security office for Rom people. Roma representatives were involved in the development of all these projects, and they consented to them (Engebrigtsen and Lidén 2010). After fifteen years, the programme had expanded and included several activities, such as free karate-instruction for girls and a youth club for Rom children and youth. The authorities and the public questioned the programme. After twenty years and a major swindle committed by the self-appointed Gypsy king's family, the programme was shut down and evaluated as a failure; two Roma had completed secondary school. No adults were wage earners, and all families received social benefits. Most families still travelled for economic and social reasons. Thus, the Roma were left on their own, that is, to the general social security system, to majority classes in schools, and to majority kindergartens. The failure of the rehabilitation project was attributed to a lack of cultural awareness by municipal authorities and the strong resistance to change among leading Roma. The outcome of this programme will be discussed in a later section.

Thanks to the welfare system, and their own businesses, the economic situation among most Roma families in Norway today is rather good. Most are still illiterate; most are semi-nomadic as they always were in Norway, travelling extensively in and outside Scandinavia during spring, summer and autumn. Several families are quite wealthy, but there are also signs of increased abuse of drugs and alcohol, conflicts between families, and other social problems. Alarming numbers of Roma children are taken into custody by the Child Welfare Services and placed in Norwegian foster homes because of alleged abuse and neglect. Families are worried about the increasing demands from the authorities that they send their children to school. Many families send their children to primary school now and then, but only two Norwegian Rom completed an education after the 1990s, no one has a regular position, and all receive social benefits. In 2011, the state launched a new programme to better the situation for Norwegian Roma and to include them in mainstream society. They established special classes for young Rom adults with work training as part of the basic education.

Some young people followed this programme for a while, but in 2013 the tension between the authorities and the Roma, and between Roma families, intensified, and school attendance stopped.

In 1999, Norwegian Roma were given status as national minority (as were the Travellers, the Kvens, the Jews, and the Forest Finns) and with this status, the pressure to integrate into mainstream society increased. The position offers protection from discrimination and instructs public agencies to facilitate the development of culture and identity, but it also raises the government's expectations of cultural and economic change among Norwegian Roma. The active negotiations with public agencies demand a new form of organization. Traditionally leadership has not exceeded the extended families and 'representation' has been in the shape of 'kings' (Engebrigtsen 2007; Kaminski 1987) – more a show for the government than anything resembling democratic representation.

Before discussing the battles between authorities and different Roma populations, I would like to discuss the social and political organization of these Roma populations, as that is crucial for understanding their relations with governments.

Social organization and cultural traits

As already stated, these two groups – Romanian Romanies migrating to Norway and the Norwegian Roma – live under different socio-economic conditions, but they share some social and cultural traits that makes it possible to see them as different groups who are engaged in the same power struggle. Roma groups are generally kin-based; they practise endogamy, and they are without institutionalized leadership above the level of the extended family. This acephalous political organization is supported by a strong sense of personal autonomy and equality between Roma, gender and age being the only dividing criteria. A traditional Rom community consists of a group of related Roma, by marriage and descent. They practise endogamy and viri-locality, and every individual has a network of relatives, most important on their father's side, but also on the side of their mother. In a family, these networks ideally make up their alliances outside the domestic group. The group that make up a settlement is not necessarily fixed, but may shift as individuals and families move in and out. The extended network is thus just as important for one's community as the local group is (Engebrigtsen 2007). Most Vlach Roma populations have an internal judicial system called *kris romani*, which is based on external objective judges and consensual decision-making (Engebrigtsen 2007; Gropper 1975; Stewart 1997; Sutherland 1975). This network formation, virilocality, and the mobile *kris* institution, imply that people must travel to keep their networks together, as also discussed by Hays and Ninkova among the Ju|'hoansi of Namibia in this volume. They must travel to visit alliances such as paternal uncles and aunts, in-laws, and influential relatives. All winter, men and some women travel extensively to funerals to show respect, they may travel to attend *kris* sessions, for work, and for business, and in summer they travel to attend marriages.

Ritual separation

Many Roma groups maintain several practices of separation from the majority populations. Romanés is the mother tongue of most Roma groups all over the world and this language functions almost exclusively orally.[9] Although Roma also speak the majority language, they often have limited command of it and use Romanés, which separates them from the majority population. As few children attend school they are separated from majority children for most part of their childhood. The Norwegian Roma, as well as many Roma populations in general, do not see education as relevant for their way of life and very few children attend school regularly over a long period of time (Engebrigtsen 2007). In Romania the school system does

9 A standardized version of Romanés is set in writing by linguists, but the majority language is mostly used when written; this may be changing these days.

not serve Roma students due to stigmatization, and in many Roma groups, most adults are unemployed and generally avoid military service, which reinforces their separation from the majority populations. Thus, traditional Roma perceive there to be a constitutive difference between themselves and non-Roma or *gaze*, a derogatory term for non-Roma. This may be seen as an adaptation to a historical and contemporary stigmatization and segregation from the majority society. These practices are supported by a cosmology of ritual purity where separation between the pure and the impure aspects of the body, of groups, of actions, and of the physical and spiritual world is basic. Historical and present experiences of exclusion together with a strong sense of difference and independence from majority society concerning, language, practices and morality, confirms a strong sense of identity segregation (Mirga 1992) in many Roma populations. Political, cosmological, and social independence from majority society is, however, paired with economic dependency (Engebrigtsen 2007; Gropper 1975; Stewart 1997; Sutherland 1975). These perceptions of difference are widely held majority by populations as well (Engebrigtsen 2007; Achim 1998).

Ethnic groups or network

The Roma are often referred to as an ethnic group, a people, or a nation in public discourse, although this reference is disputed among scholars (Gay y Blasco 1999; Engebrigtsen, 2007; Stewart 1997). Brubaker (2004) questions the idea and use of the term 'group' in anthropology and particularly in combination with ethnicity. His main objection is that ethnicity is often used as a claim for power and influence by organizations and persons without support from the people they allegedly represent. His other objection is that 'groupness', the level of incorporation in a population, is not constant, but may vary with external and internal factors, where conflict is an important one. The term ethnic group exaggerates and overemphasizes a groupness that at best is ephemeral or imagined. This point can be exemplified by the heated disputes all over Europe about the definition of Romany populations. In Norway a debate has been going on in the media about the 'sigøyner' beggars from Romania. A Norwegian Rom (head of family) was interviewed and he announced that they (the Norwegian Roma) were not *sigøyner*, but Roma, and he distanced himself from any association to 'those people' from Romania. The next day an evening paper invented a new word that embraced them all, 'romfolket' (the Rom-people), and this is now the official public name for Romanian beggars, Norwegian Roma, and traveller people, regardless of how they identify themselves. The term 'the Rom people' suggests a tightly knit ethnic group with all the popular connotations of common history, territory, interests, and organization and language. When this denomination had established itself, several Norwegian Roma realized that this new 'name' lumped them together with Romanies and Romanian Roma beggars and several Roma announced publically that they actually are 'sigøyner': Gypsies.

The idea of 'the Rom people' is a fiction of sedentary imagination (Malkki 1997). The very loosely knit network organization without leadership, or at least permanent leadership outside the family, without territorial rights and with populations spread all over the world, without a written language and historiography to consolidate groupness, presents itself as a different mode of social organization than what is generally termed ethnic groups, and with different means to pursue their interests.

Strategies and tactics: Nomadic and sedentary power modes

As the history of the Roma has exemplified nomadism and sedentarism, mobility and immobility are not necessarily oppositions, but are, as also discussed by the other chapters in this volume, combined in different ways, in different times, dependent on political and economic contexts (Barth 1980; Sørbø 1985). Thus a mobile habitus that does not necessarily depend on physical mobility (Clastres 1987; Deleuze and Guattari 1986; Scott 2009), is still a drive for movement and change. Groups that depend on physical mobility for their livelihood are often without territorial rights vis a vis a state formation and are often subject to state control and governance, which they try to avoid. This is the situation of the Romany populations in this chapter. To analyse this relationship one needs concepts that can capture the relations between a state and a mobile population and between the powers available to mobile groups and those available to sedentary groups. The French sociologist de Certeau (1984) discusses these modes of power in terms of *tactics* and *strategies*. De Certeau writes about power and resistance and is interested in how 'the weak' make use of 'the strong' to create a locus of relatively autonomous action. Strategies, as manipulation of power relations are possible 'as soon as a subject with will and power can be isolated' (de Certeau 1984, 36) and this subject (a business, an army, a city, a scientific institution) can create his/her own place, and from there manage his/her interests. The ability to manipulate power relations rests on the possibility to create one's own territory that serves as a basis from where to define and relate to others. Residing in one's own place also implies control over time and offers opportunity to plan ahead, to prepare for the future, and to write history. The Romany groups I am dealing with here do not have this option – as eternal strangers in Romania and intruders in Norway they are subject to the will, regulations, and laws of the land. Without their own place they cannot protect any specific territory and call it theirs, and thus they have no basis for legitimate knowledge; without territory they develop no legitimate history (de Certeau 1984) and without a written language it is not possible to create a legitimate common past and hence 'a people'. Of course they do this locally by oral means, but de Certeau writes that orality builds on imagination and fabulation and is thus protected from the power of established order (de Certeau 1984).

Thus strategies are possible for exactly the kinds of groups that are generally called ethnic or national, which can refer to a place or a territory that they belong to and create as their own, and control it by controlling its boundaries. Moreover, the control of time is essential for conserving and defending one's 'proper place' and by fixing a common past and future to that defined place one can be 'a people' and 'a nation'.

A tactic is, contrary to a strategy, a calculated action or a manipulation of power relations 'by the absence of a proper locus' (de Certeau 1984, 37) and by the absence of the control over time. The space of tactics, says de

Certeau, is the space of the other. Thus tactics must play on and with a terrain (read rules and conditions) imposed on it and organized by the law of the 'other'. 'Tactics are thus to manoeuvre within the enemy's field of vision and within enemy territory', de Certeau (1984, 37) wrote, and continued:

> It does not, therefore, have the option of planning general strategy and viewing the adversary as a whole within a district, visible, and objectifiable space. It operates in isolated action, blow by blow. It takes advantage of 'opportunities' and depends on them, being without any base where it could stockpile its winnings, build up its position, and plan raids. What it wins it cannot keep. This nowhere gives a tactic mobility, to be sure, but a mobility that must accept the chance of offerings of the moment. It must vigilantly make use of the cracks that particular conjunctions open in the surveillance of the proprietary powers. It poaches in them. It creates surprises in them. It can be where it is last expected. It is guileful ruse.

Migrants, tourists, and EU citizens: Controlling mobility

Controlling borders is one of the state's main tasks. The new EU internal mobility of poor, unemployed, and to a certain degree unemployable people from countries hit by economic recession and from the new member states in the former Eastern Europe appears to have struck governments quite unexpectedly. Classification is crucial for the exercise of government control (Deleuze and Guattari 1986) and for the individual's rights and duties, but the new street-working Romanian beggars challenge conventional categorization. They are categorized as tourist in the migrations regime, but as they travel for economic reasons, they do not fit into this category. They are not regular labour migrants as they have no formal qualifications and they do not seek asylum or work. As EU citizens they have the right to travel anywhere in the EU as tourists. As long as they are not employed, they are technically tourists and have no national citizens' rights in Norway except for emergency health care and social assistance. Their appalling poverty in Romania and their destitution in Norway does however challenge the social-democratic government's rhetoric of poverty eradication; it raises religious leaders' and left-wing politicians' indignation, and ordinary citizens' social conscience. For the seven to eight years prior to the time of writing this chapter, spontaneous camps have popped up both in the centre of Oslo and in the woods surrounding the city. The authorities clear these camps regularly, but they always pop up somewhere else. The perpetrators are fined, but as they are not registered, they change their names or leave the country. If they break the law several times they may be expelled, but come right back. Beggars with cars are prohibited from parking overnight anywhere, but they play hide and seek with police, gathering in the evenings and driving around Oslo and spreading out for the night. The prohibition of sleeping in cars and the practice of fining people who sleep outside in the city force people to move around often several times a night. In 2012, a group erected a camp outside a church in a central part of Oslo. This developed into a demonstration, backed by Norwegian activists, for a permanent campsite. Only a small portion of the migrant Roma participated in this demonstration. Being EU citizens and tourists, their claims had no legal grounds, but it stirred a heated debate about morality and humanism. After a while, a private entrepreneur offered them a parking lot at the outskirts of the city, but the neighbours complained so much over trash and filth that the police cleared the camp, which then popped up in the woods outside Oslo. This camp was guarded by activists and was kept clean and neat, but after several months the owner demanded the property cleared, and so it continues. The Romanies have no other option than to exploit the possibilities they have, most of them illegal: 'It must vigilantly make use of the cracks that particular conjunctions open in the surveillance of the proprietary powers' (de Certeau 1984, 37). Similar incidents are happening all over the country and in the city centre. 'Clean the city brigades' and local authorities have even hired Romanian police officers to trace beggars who gather, sleep, and eat in parks, parking lots, and corners. The camp that was introduced earlier in this chapter was closed soon after my observations, just to pop up somewhere else. The Roma leave when they are asked to and return when the guards have left. The authorities have continuous discussions about banning street begging, but for a social-democratic government this has been problematic. The solution of the new right wing government is that every local council shall decide whether or not to ban begging in their community. This is a compromise, as even in the right wing coalition there are conflicting opinions about the 'Rom-people' (*romfolket*).[10] Without legitimate claims other than staying and begging as long as begging is legal and tolerated, Romanian beggars cannot plan and develop strategies to counteract the control and harassment they are subject to. They must keep their heads low and make themselves visible as beggars, but invisible as troublemakers; they operate on the space of others and must manoeuvre within that field. Thus, they make use of the moment and develop tactics that exploit the possibilities that present themselves, constantly looking out for new opportunities.

The Norwegian Roma, on their side, are struggling to keep their autonomy by being invisible and by evading public institutions such as schools, military service, and taxation. Such tactics are not new, but an inherent consequence of their historical position as stigmatized others and their response of segregation and mutual avoidance (Mirga 1992). They find their way by 'clever tricks' within the order established by 'the strong' (de Certeau 1984, 40). The Norwegian Rom have evaded control by eagerly grasping all pastoral projects (Foucault in Goulder, 2007) intended to change and improve them and their way of life and, invisibly at first, by transforming them to their own purpose. The act that initiated the programme, Rehabilitation of the Norwegian Gypsies, in the 1960s was a letter from the 'Gypsy king' to the Norwegian king in 1965, demanding assistance to be integrated into Norwegian society. The programme had the full support from the beneficiaries, but it turned out only to be by appearance. By *manoeuvring within the enemy's field*, the Roma managed to keep the recourses bestowed on them from the state without giving up their values and interests. For instance by admitting that their children need schooling, but never sending them long enough to learn, they consent and avoid at the same time. Norwegian Roma sometimes explain school avoidance by their cultural traditions of mobility and their fear of their daughters being abducted or violated. They have gladly accepted the houses build for them by the state and not, or only partly, paid rent or maintenance,

10 A majority of homeless EU migrants in Oslo are East Europeans, but the share of Spaniards and Portuguese is increasing. In Bergen, Norway's second largest city, South Europeans constitute by far the largest group.

and kept their caravans and travels. By avoiding wage labour in any form, not accessible to illiterates anyway, by attending adult education in Norwegian, but never actually learning to read, they *avoid by consenting*. By claiming social benefits and presenting themselves as clients, they escape taxation on their real income. Exactly because the Roma have few legitimate claims, their tactic is to consent in principle and reject in practice. When pressures and demands from the authorities become too much, families move abroad and continue their activities there.

Then, in 1999, the Norwegian Roma were offered a position as national minority, that was an acknowledgement of their belonging in Norway since the late 1800s, and an acknowledgement of their community as an ethnic minority with certain rights, but not of rights to a specific territory. The responses to this position is not clear, but one result seems to be an increasing conflict over representation and most families evading state control by leaving Norway to live abroad for periods.

Mobility: Creating and recreating the social

Mobility in terms of physical and mental movement is at the basis of culture, society, and habitus among the Romanies, it is what determines their strategic position and their instruments for control avoidance. However, it is exactly by this mobility that they try to secure their core relations and what they perceive as unchangeable – their families and networks, their language, and their way of living. Thus, mobility is basic for the creation and recreation the Rom life. Through moving, people in network societies keep and develop their social relations. This illustrates the interconnectedness of mobility and immobility, of change and stasis. Throughout history both the Norwegian and the Romanian Roma I am discussing here have been sedentary and nomadic, but mobility in some sense has always been a basic trait of their economic adaptation, their social world, and their habitus. The combination of forming social alliances with the majority society and still securing cultural and political independence is a defining trait of many Roma populations (and of several others, see, for example, Chapter 1 in this volume).

As migrants in Norway, the Romanian Roma engage in social relations with non-Roma, mostly as beggars. The ambiguous relationship between giver and receiver seems to spur sympathy and care in many Norwegians and to the Roma such relationships are crucial for their economic success. To establish relationships with store owners and receive leftovers at the end of the day, to have allies in restaurants who let you use the toilets, to receive alms regularly and to be able to establish a friendship with a Norwegian is part of the Roma's adaptation tactics in a strange society. While these relationships are crucial to all newcomers to a country, what makes this situation concerning the Roma interesting is that it contrasts with their general perception of being 'eternal strangers' and secluded from mainstream society.

Tactics of dichotomization and complementarization

De Certeau's (1984) conceptual universe of tactics and strategies is based on metaphors of war and violence; it is guerrilla war against the national army, parallel to Deleuze and Guattari's (1986) concepts of state and war-machine. As there is no aspect of the Rom worlds that is not marked by the non-Roma, adaptation to the non-Roma norms is their only path to survival. As much as the concepts of tactics, strategies, and war-machines still capture important aspects of the Roma/government relationship, they only seem to focus on acts of dichotomization and miss out the negotiations and the complementarization (Eidheim 1971) that form a crucial part of the relationship. The Norwegian anthropologist Eidheim introduced these concepts in his analysis of the Saami struggle for ethnic legitimacy with the Norwegian state. This struggle was about territorial and cultural rights and was different from the one I am discussing here. The Saami struggled for ethnic recognition based on historical rights to land and resources and they claimed the right to self-governance based on these territorial roots.

The Saami successfully argued for rights to territory, common history, and language. The Roma groups I have discussed here cannot claim rights to any territory – they are strangers everywhere. This does not imply that they feel like strangers, only that they cannot exert the same type of strategic power that comes with a proper place and control of time. Groups that reside on enemy territory and want to continue to reside there must be tolerated and they must present themselves as tolerable. Complying in order to reject without breaking the relationship is vital in the struggle for tolerance to people without territory (Mirga 1992). The combination of the guerrilla-metaphor and that of negotiation may shed new light over these relations. Even the 'battles' between Norwegian Roma and authorities, imply both dichotomization and complementarization.

The Norwegian Rom 'king' approaching the Norwegian king with a request for support was an act of complementarizsation; the acknowledgement of structural similarity and the will to co-operate. This act of complementarization set the stage for the programme the king requested. The rampage and abandonment of the houses were acts of dichotomization – not planned, but tactical expressions of sedentary frustration. When parents consented to send their children to school they complementarized, when they kept them at home they dichotomized. These seemingly inconsistent and ambiguous responses were utterly confusing to government institutions and supported their general understanding of Roma as immature and insecure. While the 'complementarism' of such acts was taken by the majority at face value – as genuine acts of co-operation and of having common interests – the withdrawal was interpreted as the outcome of some family heads' domination. By interpreting the acts of complementarization as a tactic that is interdependent with dichotomization, the dynamics of this power struggle is obvious.

Thus the Norwegian Roma complementarize by complying verbally to pastoral projects offered to them, and dichotomize by avoiding them in practice. This apparent ambiguous behaviour infuses ambivalence in the majority – they are both fascinated and repelled.

Romanian beggars are not subject to the same pastoral power from government bodies, but they are subject to control, surveillance and all kinds of regulations concerning their whereabouts and movements. Their mobility renders them vulnerable as 'strangers but it also protects them from control. They consent to the control they are subject to verbally, but break the regulations when necessary. According to police records, petty crime is widespread among these groups, acts that lead to rejection and mistrust among Norwegians. Many Romanian Roma and Romanies, however, especially women, actively try to establish relationships with majority persons and many succeed. These are acts of complementarization, which to some degree counteract the crime committed by others. They generally also praise the Norwegian kindness, and downplay the harassment they are subject to.

Conclusion

My objective with this chapter was to discuss how a perspective of mobility and nomadism may open up interesting interpretations of power relations between minorities, the majority public, and the state. Here with an empirical focus on two different mobile Roma/Romany populations and state authorities in Norway. Both the Norwegian Roma and the Romanian Romanies in Norway are in a position where their true interests as mobile subjects are more or less illegitimate, or seen as a threat to authorities, and cannot be presented as claims in open negotiations with them. To follow their interests they have no other choice than to play along and exploit cracks in the state's armour (de Certeau 1984). By tactics of complementarization in order to be tolerated, and of dichotomization in order to affirm their separateness, they are able to retain their livelihood and community without losing what they see as their independence.

Mobility is the core aspect of the uneasy relationship between these Roma/Romany populations and the state. Physical mobility is paired with a nomadic identity in Deleuze and Guattari's (1986) sense, which render these populations in opposition to state dominance and control. They are residing if not on enemy territory, then not on their proper territory. They are under a continuous regime of pastoral power and surveillance, and subject to programmes for cultural change or politics of rejection. Without legitimate interests, they have no legitimate claims and cannot, as ethnic groups with territorial roots, develop strategies to negotiate with the majority. The aspect of time is also important in this relationship – as mobile subjects without control over a territory, the Roma must always adapt to majority mobility regimes and strategies of placement and displacement. However, their physical mobility and unpredictability also render them difficult to control. Now they are here, then they are there, and suddenly they are gone.

I am not claiming that the Roma's use of tactics such as the combination of dichotomization and complementarization are always conscious acts – on the contrary, I assume that it is foremost an effect of their strategic situation and mobile habitus. Trust and mistrust, ambiguity and uncertainty concerning individuals' feelings and interests in relation to the majority are also aspects of the Rom/Romany/majority relationship that create emotional bonds of attraction and repulsion. This ambiguity plays on and enables mobile tactics to parry majority power, but never to conquer it.

Bibliography

Achim, Viorel. 1998. *The Roma in Romanian history*. Budapest: CEUPress.

Andriaenssens, Stef, and Jef Hendrickx. 2011. 'Street-level Informal Activities: Estimating the Yield of Begging in Brussels'. *Urban Studies*, 48 (1): 23–40.

Barth, Fredrik. 1980. *Nomads of South Persia: The Basseri Tribe of Khamseh Confederacy*. Oslo: Universitetsforlaget.

Bauman, Zygmunt. 1991. *Modernity and Ambivalence*. Cambridge: Polity Press.

Brubaker, Rogers. 2004. *Ethnicity Without Groups*. Cambridge: Harvard University Press.

Certeau, Michel de. 1984. *The Practice of Everyday Life*. Berkeley: University of California Press.

Clastres, Pierre. 1987. *Society Against the State*. New York: Urizen.

Deleuze, Gilles, and Félix Guattari. 1986. *Nomadology: The war Machine*. NY: University of Minnesota Press.

——1987. *A Thousand Plateaus: Capitalism and Schizophrenia.* Minneapolis: University of Minnesota Press.

Eidheim, Harald. 1971. *Aspects of the Lappish Minority Situation.* Oslo, Bergen and Tromsø: Universitetsforlaget.

Engebrigtsen, Ada I. 2007. *Exploring Gypsiness. Power Exchange and Interdependnece in a Transylvanian Village.* London and New York: Berghahn Books.

——2012. Tiggerbander og Kriminelle Bakmenn Eller Fattige EU-borgere? Myter og Realiteter om Utenlandske Tiggere i Oslo. *NOVA Notat* 2/12. Oslo: Norsk institutt for forskning om oppvekst, velferd og aldring.

Engebrigtsen, Ada I., and Hilde Lidén. 2010. 'Å finne sin plass som minoritet – Rombefolkningen i Norge i dag'. In *Nasjonale Minoriteter i det Flerkulturelle Norge,* edited by Anne Bonnevie Lund and Bente Bolme Moen, 199–211. Trondheim: Tapir akademiske Forlag.

Engebrigtsen, Ada. I., Johanna Fraenkel and Daniel Popet. 2014. 'Gateliv. Kartlegging av Sittuasjonen til Utenlandske Personer som Tigger'. *NOVA Rapport 7/14.*

Gay y Blasco, Paloma. 1999. *Gypsies in Madrid. Sex, Gender and the Performance of Identity.* Oxford: Berg Publishers.

Glick Schiller, Nina, and N. B. Salazar. 2013. 'Regimes of mobility across the globe'. *Journal of ethnic and migration studies,* 39 (2), 183–200.

Golder, Ben. 2007. 'Foucault and the Genealogy of Pastoral Power'. *Radical Philosophy Review,* 10 (2): 157–76.

Gropper, Rena C. 1975. *Gypsies in the City: Culture Patterns and Survival.* Princeton: The Darwin Press.

Halvorsen, Rune. 2000. 'Etnisk Mobilisering i en Brytningstid'. In *Romanifolket og det Norske Samfunnet. Følgende av Hundre Års Politikk for en Nasjonal Minoritet,* edited by Bjørn Hvinden, 195–226. Bergen: Fagbokforlaget.

Hanisch, Ted. 1976. *Om sigøynerspørsmålet: en undersøkelse for bakgrunnen for sosial konfrontasjon.* Oslo: Institutt for Samfunnsforskning.

Hvinden, Bjørn. 2000. 'Innledning: Fra Fordømmelse til Respekt og Verdighet'. In *Romanifolket og det Norske Samfunnet. Følgende av Hundre Års Politikk for en Nasjonal Minoritet,* edited by Bjørn Hvinden, 11–31. Bergen: Fagbokforlaget.

Johansen, Jahn Otto. 1989. *Sigøynernes Holocaust.* Oslo: Cappelen.

Kaminski, Ignacy-Marek. 1987. 'The Dilemma of Power: Internal and External Leadership. The Gypsy Roma of Poland'. In *The Other Nomads. Peripatetic Minorities in Cross-Cultural Perspective,* edited by Aparna Rao, 323–56. Köln and Wien: Bölaug Verlag.

Lidén, Hilde, and Ada I. Engebrigtsen. 2010. 'De Norske Rom - og Deres Historie'.
In *Nasjonale Minoriteter i det Flerkulturelle Norge,* edited by Anne Bonnevie Lund and Bente Bolme Moen, 87–96. Trondheim: Tapir akademiske Forlag.

Malkki, Liisa H. 1997. 'National Geographic: The Rooting of Peoples and the Territorialization of National Identity Among Scholars and Refugees'. In *Culture, Power, Place: Explorations in Critical Anthropology*, edited by Akhil Gupta and James Ferguson, 52–74. Durham, NC: Duke University Press.

Mirga, Andrzej. 1992. 'Roma Territorial Behaviour and State Policy: The Case of the Socialist Countries of East Europe'. In *Mobility and Territoriality. Social and Spatial Boundaries Among Foragers, Fishers, Pastoralists and Peripatetics*, edited by Michael J. Casimir and Aparnu Raoand, 259–87. Oxford and NY: Berg.

Pettersen. Karen-Sofie 2005. *Tatere og Misjonen: Mangfold, Makt og Mostand*. NOVA-rapport 2/2005. Oslo: Norsk institutt for forskning om oppvekst, velferd og aldring.

Rosvoll, Maria. 2013. 'Antisiganisme'. *Materialisten*, 40 (1–2): 7–39.

Salazar, Noel B., and Alan Smart. 2012. 'Anthropological Takes on (Im)Mobility'. *Identities: Global Studies in Culture and Power*, 18 (6): ix. Accessed: doi: 10.1080/1070289x.2012.683674.

Scott, James C. 1990. *Domination and the Arts of Resistance: Hidden Transcripts*. New Haven and London: Yale University Press.

——2009. *The art of not Being Governed: An Anarchist History of Upland Southeast Asia*. New Haven: Yale University Press.

Sørbø, Gunnar M. 1985. *Tenants and nomads in Eastern Sudan: A Study of Economic Adaptation in the New Halfa Scheme*. Uppsala: Nordic Africa Institute.

Stewart, Michael. 1997. *The Time of the Gypsies*. Colorado: Westview Press.

Sutherland, Anne. 1975. *Gypsies, the Hidden Americans*. New York: The Free Press.

Urry, John. 2014. *Offshoring*. Cambridge: Polity Press.

KJERSTI LARSEN

3 'This Is Where We Belong': Migration and Intersecting Mobilities in Zanzibar Town, Zanzibar

Discussing mobility and belonging, societies within the Indian Ocean region and along the east African coast appear almost emblematic. Historically, spatial mobility there has been interlinked with social and economic processes and the formation of societies (Sheriff 2014, 6). Facilitated by the monsoon, the East African coast formed part of the regions commercial system for 2,000 years, 'at least, as an intermediate zone of exchange' (Sheriff 1987, 8). Moreover, historians have documented, for instance, how between 1500 and 1960, oceanic connections, commerce, and trade were further intensified and interlinked with a variety of temporary or permanent movements of people and the exchange of ideas, knowledge, social, and cultural practices as well as commodities (Alpers 2000; Gilbert 2002; Horton 1990; Horton and Middleton 2000; Sheriff 1987, 1992, 2008, 2010). Throughout this period new forms of extensive trade and sociocultural interconnections were characterized by particular economic, political, ideological, and technological contexts; these included Islam, the emergence of coastal communities and small port cities around the ocean engaging in maritime trade and, later on, the British imperial power and long distance trade dominated by sailing crafts and, eventually, steamers. A body of scholarship demonstrates how transnational connections at the political, economic, and cultural level have played a constitutive role in the formation of local life-worlds and societies within the Indian Ocean region and along the east African coast, including Zanzibar (Bishara 2014; Campel 1989, 2005; Pearson 2003), as well as perceptions of identity and belonging (Gurnah 1997, 2008; Topan 1991, 2003a, 2003b, 2004). Thus historically, on Zanzibar and, more generally, in the Indian Ocean region not only movement and fixity, but also connectivity and multiple belongings are historically part of a complex system of adaptation (Sheriff 2010, 2014) and yet have a specific contemporary configuration.

In this chapter, current perceptions and experiences related to connectivity and intersecting mobilities in Zanzibar Town are explored, and the ways in which cultural configurations combined with changing social, political, and economic circumstances may condition notions as well as modes of movement and fixity will be analysed.[1] Attention is here paid to the manner in which perceptions and practices of movement and interconnection are integral aspects of past and present, with the future in mind. Discussing mobility, I am concerned with how actual and imagined processes of locations and relocations are lived, and how political and economic change may influence the conceptualizations of life trajectories, identities, and relationships. Regarding people's life trajectories and recollection of what precisely affected the decisions made would, I argue, be fashioned by what Makris (1996, 175) has described as 'the "lived in" political conditions of everyday life' (175).

Following this, and comparable to the other chapters in this volume, the focus will be on mobility seen from 'below' and within a particular cultural and socio-political framework. How migration has been perceived and practised by ordinary Zanzibari women and men or households and, with time, in what way shared experiences actually fashion peoples' future prospects. It concerns not only how migration experiences become part of a particular society's collective memory, but also the ways in which a perception of mobility is intertwined with local socio-political and cultural circumstances which over time and relative to gender, age and social position, affect people differently. Thus, an analysis of mobility concerns the movement of people both in terms of temporal and spatial dimensions. Gender is a significant feature in the organization of society as well as the in the formation of social personhood. This means that also the potentiality and trajectory of women and men's social and spatial

1 Since 1964, Zanzibar together with Tanganyika formed a federation named Tanzania. Zanzibar consists of two main islands, that is, Unguja and Pemba and some small islets like Tumbatu. The capital of Zanzibar is Zanzibar Town.

mobility would differ not only between societies, but also within the same society. Interesting in the context of Zanzibar Town is thus both in what manner women and men perceive and live their lives and identity in terms of movement, localization, and belonging and to what extent mobility could be said to form part of what they perceive as social order. Could mobility be seen as constitutive of society, that is, as internal to social organization and to identity formation? Recently, Freitag and von Oppen (2010) have argued that the concept 'translocality' could be helpful when addressing 'tensions between movement and order' and, moreover, regarding how 'to cope with transgression and with the need for localizing … order' (8). Through ethnography, I wish to explore whether an approach focusing on 'mobility' would bring to the fore an understanding of spatial movement as embedded in a certain mode as well as modality of life that has emerged over time and become part of a socially shared memory (see Chapter 1 in this volume). Contrary to what is usually associated with 'migration', that is, political crisis and potential social conflicts in the wake of either labour migration (Mitchell 1956, Ferguson 1990), forced migration (Harir 1994, Indra 1999) or refugeeism (Bascom 1998, Black and Koser 1999), an approach emphasizing mobility would mean that there should be no tension between movement and order.

Before discussing experiences and practices of mobility and belonging in contemporary Zanzibar, a brief comment on the legacy of the 1964 Revolution, especially regarding movement of people and commodities, is needed. The 1964 Revolution, which took place shortly after Zanzibar gained independence in 1963, overthrew a democratically elected government (Cameron 2004, 105). It totally changed the political order and state system and in turn reshuffled the previous social structure and thereby also the permissible and economic status of many households, families and individuals. The aim of the revolutionary leaders was to overthrow what they defined as a slave holding-society and feudal Arab state privileging inhabitants of Arab, Indian and Comorian origin at the expense of the unprivileged African majority population. The revolution was voiced in terms of 'ethnic essentialism' (Cameron 2004, Glassman 2011) and society was represented in ways that legitimated the violent revolution in terms of questions of power, politics of inclusion and exclusion and the right to belong (Larsen 2004). During the upheaval as well as in its aftermath, many women, men, and children were killed, while some families managed to flee or were forced to leave. It turned a number of women and men into refugees and migrants. Moreover, in the years afterwards, many so-called non-Africans according to the dominant identity-labels at the time, especially Zanzibaris of Asian, Comorian, and Arabic descent, who had survived the days of the revolution and who thereafter stayed on, time and again felt threatened by the introduction of new polices such as the political decisions of 1967. According to these, all Zanzibaris of Asian descent were to be expelled from civil service positions. Then, in 1971, with the new regulation of business licenses and ownership issued, many so-called non-African lost their basic livelihoods as owners of small shops and workshops.

Important for the discussion here is that the Revolutionary Government, which came into power following the 1964 Revolution, also introduced a restrictive policy regarding international communication and connections – travel permits and passports became difficult to acquire. From what people I worked with have narrated, only trusted members of the government or government employees could, from time to time, legally travel abroad. National visa regulations were indeed strict, and foreigners who were allowed into the islands were under strict surveillance. Notwithstanding, people found ways to leave the island-state. However, then as now there was a gender dimensions to this mobility. While married couples and families would, if possible, migrate together and to places where they knew they would find relatives, or at least other Zanzibaris, young men would try to leave on their own or in smaller groups. Women would seldom migrate alone and those who did so would usually go on to live with relatives living abroad.[2]

Despite the explicit racial dimension of the 1964 Revolution and the governance of the new state dividing the population according to crude ethnic categories, Zanzibar, like other societies within the Indian Ocean,

2 This does not mean that women cannot also travel alone. Adult and elderly women could travel alone or in the company of other women, for business or when together with their children, visiting relatives.

nonetheless continues to encompass perceptions of human diversity whether linked to gender, *kabila* ('tribe' or ancestral place of origin), or rank (Eastman 1971; Giles 1987; Larsen 1990, 1998a, 2008; Middleton 1992; Prins 1961). A body of literature discusses how this diversity, which is linked to a long-term history of spatial and social mobility, is also characteristic of the shared routines of everyday life and practice (Fair 2001; Larsen 2000, 2004; see also Simpson and Kresse 2008). This understanding incorporates the idea that movement does not necessarily have to 'involve a denial of collective identity' (Allen and Turton 1996, 11). As I have discussed elsewhere regarding Zanzibar Town, over time particular configurations of place, motion, and belonging have become incorporated on the level of individual identity and into the life histories of numerous individuals, households, communities, as well as society as such (Larsen 1998a, 1998b, 2004, 2008, 2009).

Methodological and ethnographic considerations

In this chapter I am drawing on data collected during longitudinal ethnographic fieldwork in Zanzibar Town, from 1984 until the present, and moor the discussion in the local reality by focusing on two households with which I have had long-term relations. This approach allows me to discuss how particular social parameters together with changing political and economic policies may fashion people's lives and reorganize social relationships. The discussion will touch on the phenomenology of mobility and people's concerns and perceptions regarding their local, trans-local, and international experiences. Focusing on the urban setting of Zanzibar Town, it is, however, important to keep in mind that a particular form of cognitive and spatial flexibility is already incorporated into people's perceptions of identity and place. As already mentioned, this society incorporates perceptions of human diversity in which a notion of movement forms an intrinsic component. In terms of belonging, people tend to locate themselves by referring to both Zanzibar and to one or several places of origin (*ya asili*) elsewhere. In addition to being a Zanzibari with a Tanzanian passport, people would refer to themselves as 'persons of' for instance Yemeni, Indian, Baluch, Makonde, Comorian, or Arabian origin. Usually people refer to what they call *kabila*, translated into English as 'tribe'. This term, I have argued elsewhere, mainly denotes places of origin beyond Zanzibar[3] (Larsen 1998a, 1998b, 2004). The presence of such categories of identities and places of ancestral origin implies that an idea of movement is already incorporated in their perception of belonging and being (Larsen 2008). Such a conceptualization of identity exceeds the dichotomy between mobility and immobility in the sense that a person is from Zanzibar, yet may originate from 'elsewhere'. It reverberates what Gilbert (2002) has rightly suggested: 'Cultural identities are less a product of origins, than they are the product of shifting and contingent historical forces' (32; see also Chapter 1 in this volume) and, I would add, social relations.

The women and men I have come to know during fieldwork have all directly or indirectly 'lived' with the legacy of the revolution and its political and economic aftermath (Larsen forthcoming). Important here is that the post-revolutionary state officially launched an identity politics where, officially, all Zanzibaris were to be considered as Afro-Shirazi. The Forced Marriage Act, which was part of this policy, allowed men of so-called African descent to marry women of Arabian, Indian, and Comorian descent without the women's families consent. The idea was that such 'mixed' marriages could further eradicate the meaning of *kabila* so that all Zanzibaris would eventually 'have to' perceive themselves as being Afro-Shirazi. Yet, even this new, politically invented identity label did indeed refer to two different places of origin (*asili*), that is, Shiraz in Persia and Africa, and as such it implied an acceptance of historical movements and connections within the region – of multiple belonging. Ideologically, the aim was to suppress previous identity configurations. Nevertheless, until today vernacular identity categories and thus all distinctions with reference to *kabila* are still applied – these identity categorizations indicate the

3 The term may however also refer to particular categories of people such as Hadimu, who are considered indigenous Zanzibaris or Watumbatu.

historical significance of mobility and trans-local and national connections within the Indian Ocean region.[4]

Against this background, the two cases presented below would be representative of the lives and experiences of many Zanzibaris. One of the households belongs to a family of Indian descent (*Wahindi*) and the other to a family of Comorian descent (*Wangazija*). My choice of households does not imply that I concentrate the discussion on families who, for instance, in European contexts, would be considered to form part of 'immigrants' or 'diaspora communities' – rather, these are typical Zanzibar Town households precisely because they have emerged from certain historical modes of trans-local connections within the wider region. Both cases focus, more or less, on individual persons, yet the situation and choices discussed concern household dynamics at large. The first case discusses the ways in which spatial and social mobility may become incorporated into a household and how this kind of mobility may affect and become affected by identity formation and household structure over time. The second case focuses on how assumptions about an assumed linkage between social and spatial mobility are incorporated into daily life discourses on potential futures without ever being fully realized, but could still condition the lives of those involved and a household trajectory.

Mobility and connectivity

Fatima, a woman in her late seventies, was born in Zanzibar, like her mother before her, while her father arrived in Zanzibar with his parents who migrated from Gujarat in India to Zanzibar at the very end of the nineteenth century, a time when there were subsequent periods of drought

4 The invention of the 'new' identity label until then unfamiliar to people, not reverberating any form of social or individual memory, seem to have perpetuated vernacular identity perceptions on private arenas. Interestingly so, until today, the pre-revolutionary identity categories is often applied in everyday-life discourse.

and hunger in the area where they used to live. The family of her husband Yusuf are from the same area in Gujarat as well as what they call 'tribe' (*kabila*) as her own family. However, unlike her, Yusuf was born in Gujarat and arrived with a dhow in Zanzibar together with his parents and three siblings when he was a young child. Fatima and Yusuf have twelve surviving children of whom the oldest is now in her early sixties. Regarding questions of mobility, place, and identity, they all see themselves as Zanzibaris of Indian origin, that is, *Wahindi*. Yet, in common with most Zanzibaris of Indian or Asian origin, time and again, they would contend that in periods of economic decline or elections, some 'political voices' would always utter that due to their Asian ancestry they do not belong in Zanzibar. Fatima explained:

> Somehow, it is true that our place of origin is India (*kwetu India*), but we have never lived there; we have only lived here. This is where we belong. We have nowhere else, and this is where we are born and have lived. Still given the unsettled political situation in Zanzibar and being of Indian origin we can never trust that we will be politically and economically protected by our Government.

This quote expresses the extent to which mobility and notions of belonging are inherent to her approach to the human condition, to life; yet, at the same time, she has hesitations about whether future political regimes will share her understanding. Fatima's comment suggests that in terms of cultural identity she sees herself as belonging in Zanzibar. Nevertheless, her explanation also reverberates social memories of the revolution, including the politics of exclusion, formalized through specific regulations introduced in 1967, 1970, and 1971, limiting the possibilities of people of Asian and Arab descent to hold private enterprises, employment in public sector, the Forced Marriage Act, and emigration restrictions – memories which until today penetrate all political discourse in Zanzibar as well as personal narratives, especially of people who are above forty years of age.[5] These

5 The violent upheavals and the political and economic changes that followed the 1964 Revolution had manifold consequences for individuals, families and for society and the structural dynamics between society and the state (Babu 1991; Bennett 1978; Cooper 1980; Glassman 2011; Lofchie 1970; Myers 2003).

memories together with actual political and economic circumstances, which have played on notions of race and, at the same time, exposed the question of autochthony and thus the connection between migration, mobility and belonging, have fashioned people's choices and movements. Let me briefly elaborate.

In the mid-1970s, two of Fatima and Yusuf's sons and one of their daughters moved abroad. While the sons left Zanzibar for Greece illegally in order to look for employment on Greek-owned ships, the daughter Rukia moved to Pakistan with her husband just after her wedding. Some of her husband's relatives, then settled in Pakistan, had invited him to join their business. Regarding the daughter's marriage, Fatima explained, 'Back then life was difficult here and we saw this marriage as an opportunity for her to leave Zanzibar for a better future. At that time, this appeared the right thing to do'. In the period following the revolution until the mid-1980s, it was difficult to obtain emigration permits from the state, and thus this marriage provided a possibility for their daughter to legally leave Zanzibar and thereby also escape a potential forced marriage. Only after the mid-1980s, when the government adopted political and economic liberalization within the terms of what became known as the Structural Adjustment Programme, did national citizens reacquire a universal right to hold passports.

Fatima and Yusuf's sons left for Greece to look for employment on foreign ships so as to be able to generate an economic foundation with a hope for future investment at home in Zanzibar. During their stay abroad, the two sons working on Greek-owned ships were able to send remittances to their family household as well as save money for their planned future return. Following the political and economic liberalization process, which started in the mid-1980s, Fatma and Yusuf's sons did return and gradually they repatriated and invested their savings in the family business in Zanzibar. Due to the fact that their return coincided with the introduction of an economic environment favouring people with access to foreign currency on the island-state, over time the family managed to extend their business venture based on import of foreign commodities so as to include hotel-business and supermarkets (see Kheshodkar 2010, 232). Within a rather short period of time, the household's previous engagement in spatial mobility combined with a changing political climate and the introduction of new and liberal economic environment in Zanzibar further facilitated a pronounced social mobility. An interconnection of that sort, between spatial and social mobility, is not inevitable. In the past, Fatima and Yusuf's relatives had moved in the search of improved life conditions also within the Indian Ocean region. However, among them, no one seemed to have experienced any significant form of social mobility either in Zanzibar or elsewhere in the region.[6] Seen over time, it also becomes apparent that an interlinkage between spatial and social mobility may only be temporarily due to individual, societal, or political transformations. This kind of fluctuating social mobility in the context of spatial mobility was experienced by Rukia, who moved to Pakistan after her marriage.

Then, in the early 1990s, Fatima's daughter, Rukia, returned from Pakistan together with her husband, who by then had been out of work for several years. Their life abroad had become dependent upon remittances from Zanzibar. After their repatriation to Zanzibar, they divorced and the ex-husband moved on to Mombasa in Kenya, where he could join a small business run by one of his brothers. Rukia re-married a relative who had left Zanzibar just after the 1964 Revolution, and who was still living abroad. Again she left Zanzibar. Without entering into any details regarding the divorce, I wish to emphasize how the subsequent migration of Rukia as well as her first husband verifies a rather common feature of life in the region, namely, that members or branches of the same family settle in different places along the East-African coast. Historically, such trans-border connections and modes of adaptations have been common within the Swahili region[7] (see also Bang 2003; Freitag and von Oppen 2010; Walker 2010). Currently Rukia visits Zanzibar from time to time with her husband and children, and due to the social mobility of the family household, Fatima,

6 This comment relates to the history of this particular kin group. For other kin groups reaching Zanzibar from other regions in Asia, the situation would have been different. However, such a discussion cannot be included in this chapter.

7 For instance, Bang (2003) discusses how in the Indian Ocean region, scholarly networks have long provided ties and shared discourses across places and milieus, between ideas and events despite physical distance.

together with some of her other married daughters with children and whose husbands run small businesses in Zanzibar, managed to obtain visas to visit Rukia abroad. Seen in a gender perspective, what is important here is that it was once again marriage that made Rukia leave Zanzibar, while her ex-husband relocated mainly due to economic and political circumstances. Still, what Rukia and her ex-husband have in common is that both relocated within long-established trans-local and international family or kin networks. In addition to depicting how movement is incorporated into life trajectories and social relationships, this case suggests that women and men's movements are organized differently and in accordance with the dominant, locally produced gender system.

When mobility appears plausible, yet becomes impossible

As the other chapters of this volume also show, mobility forms part of encompassing cultural, social, and economic configurations. Additionally, in my aim with this chapter is to explore how perceptions and practices of mobility and interconnection evolve with time. Seen from the household of Fatima and Yusuf, complex inter-linkages between mobility and immobility, local society and transnational connection, as well as perception of rootedness come to the fore. All these aspects, in various ways, form part of people's life trajectories and become ingrained in perceptions of mobility. Everyday practice and discourse in Zanzibar Town often evolve, precisely, around prospects and ideas of migration and how to improve one's social standing – although the actual mobility of many individual persons and households may be rather restricted. Regarding relocation, even travel to the mainland or within the Indian Ocean region would be considered too costly by many. And, if by any chance, assets should be available to purchase a return ticket for transnational travel and access to means necessary to secure a stay abroad for a certain period of time can be documented, a visa to a place like Europe might still be impossible to obtain. In general, people often refer to rumours claiming that the visa sections at various embassies regard plans for international travel to be disguised strategies for migration abroad without any return-prospects. Obviously, in such settings people's experiences merge with the narratives of others who have been denied visas to foreign countries, in particular, countries of the 'North'. The experiences, attitudes, policies, and regulations referred to in these discussions stand in stark contrast to current and abstract visions of people's potential movements which tend to refer to what seem for some to be 'an ever-increasing mobility and cross-border communication in a globalizing world'. Rather, what appears as the actual situation on the ground for, in this particular case, women and men in Zanzibar is restricted social and spatial mobility.[8] A wish for a better or even different life elsewhere often involves economic aspirations, yet, as will be explored through the following case, other complex, locally configured social mechanisms may also be involved. With this intermediate comment, I now turn to the second case.

Feeling stuck in a world of envisioned mobility

The life trajectories of many Zanzibari women and men are moulded by visions and practices related to both social and spatial (im)mobility. Their plans, choices, actions, and imaginations usually imply some form of movement whether social, spatial or both. Aisha is a woman of Comorian descent (*Mngazija*) whom I have known since 1985. She is now in her early fifties. Her maternal grandmother came from the Comoros (*Ngazija*) to Zanzibar together with her parents when she was still a young child. Some of their relatives still live in the Comoros, and some continue to travel between the two places. Several other relatives live abroad in Europe, Canada, and the US. They left Zanzibar as asylum-seekers in the wake of violent multi-party

8 This situation they share with most people from the so-called 'south' or developing countries, especially if one does not belong to the multi-local elites. By 'multi-local elites' I allude to those with access to and control over resources as well as the ability to engage in correlating activities or undertakings across state boundaries.

elections that took place, particularly during the period 1995–2005. Among Aisha's immediate family only she and two of her sisters, one married and one divorced, have always lived in Zanzibar. One brother and one sister left as asylum-seekers after the 2005 election. Due to unemployment, economic problems and thus downwards social mobility abroad, the brother has now returned to Tanzania where he is currently doing business from Dar es Salaam, while the sister has remained abroad together with her Zanzibari husband. The husband of the other married sister still living in Zanzibar is a labour migrant in Dubai, and for many years the divorced sister has engaged in business and travelled between Dubai and Zanzibar. For long periods Aisha's mother has resided in Dar es Salaam, where she owns a small apartment building with a ground floor with shops that she rents out. However, time and again, Aisha's mother emphasized that Zanzibar is home; it is where she belongs.[9] For instance, this means that she would always make a point of preparing for and spending Ramadan in Zanzibar. Ramadan, the Muslim fast, is here perceived as the period where society and one's belonging is socially marked (Larsen 2011).

As already mentioned, Aisha has several relatives and friends living abroad, and she remains connected to them, mostly through the phone. Aisha told me, 'Daily we exchange news, photos, and small videos of our children or, simply, of the food we have prepared, our new outfits and styles, as well as the various events taking place. Recently we have also started daily communication through the free *WhatsApp* net'. Thus access to means of communication secures a kind of localized 'everyday presence' also of those living abroad and vice versa. People's use of the new media intensifies this form of connectedness. Daily conversations and information criss-cross between households, over neighbourhoods, as well continents in such a way

9 A friend of Aisha lives another kind of mobile livelihood, which I cannot discuss in this chapter. Wahida is a Zanzibari from the island of Pemba and an Omani citizen. Her family is of Arabian decent (*Waarabu*). In the wake of the 1964 Revolution they left Zanzibar to live in Oman. During recent years, Wahida, who, like many others in the same situation with relatives still in Zanzibar, has been travelling between Oman and Zanzibar. Currently, after having been reinstated as a Zanzibari citizen, she has been able to build a house just outside Zanzibar Town.

that they form a kind of virtually shared moral space, in spite of physical distance and places. As such, there is a new degree of interconnectedness between Zanzibar and especially Zanzibaris living elsewhere. This situation creates a sense of shared moral universe that is independent of physical place and presence – at times, ignoring the economic, political, and ideological parameters that circumscribe the different livelihoods involved. This may further intensify an idea of Zanzibar as a main reference for 'home' and thus as 'the moral community' also in milieus elsewhere.[10] Within this moral community gender is marked, but also makes a difference. This does not mean that people within each of the gender categories, share exactly the same aspirations or behave in the same way. Still, the dominant gender system does produce a social framework that fashions women's as and men's lives in different yet interrelated ways. Thus gender as a social category also affects their understandings of and approaches towards mobility, a point that is also noted by Simonsen in his chapter on perceptions of mobility among the Mambwe in Zambia (see Chapter 6 in this volume).

Next to being in communication with people living elsewhere and in different economic, political, and ideological conditions, Aisha has also searched the outside world as it has materialized itself through persons and relations in Zanzibar. Over the years Aisha met and formed relationships with foreigners who have lived and worked in Zanzibar for periods of different durations. Contrary to Rukia who has married and moved abroad together with Zanzibari husbands, Aisha has always assumed that she would marry a 'European' man, *Mzungu*, not a Zanzibari. Aisha, like Rukia, has, however, known that her possibility for spatial as well as social mobility would depend on a relationship with a man, ideally through marriage. Such an expectation would be in line with dominant cultural values. For a period Aisha was married to a European working for a business firm in Zanzibar, with whom she has no children. After they divorced, she lived for several years with another European man working for an international

10 Communication through the Internet and mobile phones has, in principle, become highly accessible. However, given the quality of the educational system there are many Zanzibaris who have difficulties in producing and relating to written texts not only in English but also in Swahili.

development aid agency, with whom she has one son. When this latter man's working contract expired, he, like her first husband, left Zanzibar. Aisha's relationships with the above-mentioned men, who have both had temporary residence in Zanzibar, have been intertwined with hopes for long-lasting social and spatial mobility. Such hopes are not based on pure fantasy, but carved from observations and rumours concerning other women who have been married to foreigners. Aisha knew several who had been able to move abroad due to a marriage with a European (*mzungu*). After some time, some of these women have returned and re-settled in Zanzibar as women of means. Among their common friends, narratives about these women whose life trajectory they know, are time and again evoked in daily conversations where people share stories about their own mobility experiences as well as those of others – the fortunate as well as the unfortunate.

Aisha has travelled within the region, and thus not remained immobile.[11] None of her travels has so far resulted in social mobility. Having travelled within the region, she has, nevertheless, always expressed a wish to go to Europe (*Ulaya*) as so many others in her close surroundings have already done. 'Not necessarily to live there, but to see and to have an experience of what life in Europe is like' she often told me. Repeatedly, she has mentioned that it is almost embarrassing that she, who has come to know numerous Europeans (*Wazungu*), has never had this opportunity. In her view, a journey to Europe would have a significant effect on her esteem among relatives and friends. She contended:

> It would do me good, at least, to have seen *Ulaya* [Europe] and how life is there (*vipi kuishi kizungu*) with my own eyes. I feel that people do not take me seriously because I have only observed and been with *Wazungu* here in Zanzibar.

Yet, at other times, Aisha would say that her wish to travel is mainly for economic and social reasons.[12] This is linked to how Aisha, like many other

11 The term '*Ulaya*' means Europe, but is often used to refer to the broad category 'the western world'. Likewise, the term *Wazungu* is used for 'westerners'.

12 Her younger brother, who left Zanzibar after the 2000 elections and settled in Europe for a while, has told her that even though he had to accept menial work in a supermarket, he was able to earn a decent living. Commenting on her brother's

young and adult Zanzibari women and men, does not have a regular income. In addition to high unemployment, those who went to school between the 1964 Revolution and the early 1990s, in particular, suffer from erratic formal education and training, which easily block their possibilities for any work in the civil service. This being so, they nevertheless continuously plan different kinds of business ventures. What these projects have in common is that they all involve transnational movements and connections. As such, for many whom I have worked with, their immediate associations to wealth and economic success, more or less consciously, seem to be embedded in historical narratives about long distance trade within the Indian-Ocean region, linking spatial and social mobility. Whenever we have discussed the possibility of travelling abroad, the economic dimension has played a significant role, although it is never the only issue associated with travel. In situations when Aisha has referred to the travels she had previously conducted within the region, these were always presented as 'leisure-travels' and as invitations from friends and relatives in Kenya, Oman, and, more recently, in South Africa. However, when she has talked about future possibilities, for instance, to commute between Zanzibar and the Comoros, her plans are linked to a business venture. She would say, 'If I only had the means to go to the Comoros ... I could start some business of my own with vanilla'. As mentioned above, one of her sisters has for many years engaged in business between Zanzibar and Dubai – an arrangement that has secured a relatively stable livelihood. This was made possible due to her own persisting work and because a relative in Dubai was able to provide the initial funds necessary for the start-up of the business-enterprise.

Seen in a time-perspective, Aisha recently conveyed that she felt more and more caught in her life by her previous choices and the social expectations that she herself has endorsed. In particular, she regretted that she had been made to believe that social mobility could be achieved through marriage with a 'European' (*Mzungu*). She criticized the men she previously met and co-habited with for only wanting to be with her while in Zanzibar, leaving her without any future prospect of economic or

experience, Aisha would say, 'At least, I would have my salary, even if work would be hard and of low rank'.

social certainty. I would say that she has rather been left in the shadow of their transnational livelihoods. Her possibility to secure social and spatial mobility through a relationship with a European man, due to her aging, is slowly decreasing and, at times, she has expressed disillusion concerning the future. Interestingly, Aisha argued that she has never felt criticized because of her relationships to European men, but only because she has not been able to convert these relationships into any consequential social mobility. With time, it seems that Aisha no longer sees any kind of mobility as an option for herself. Rather, she would time and again express a growing feeling of immobility while, nevertheless, continue to try to locate new opportunities. To this extent, Aisha's attitude would be in line with what Salazar (2012) calls 'a cultural logic of migration' (18).[13] Methodologically, longitudinal studies provide access to peoples' life trajectories, which, in turn, reveal the manner in which experiences and ideas regarding fixity and mobility may shift with time as shown by the case of Fatima's family as well as by the life trajectories of Rukia and Aisha discussed above. Feelings of being stuck and notions regarding plausible mobility are clearly interconnected with social facets such as living conditions, age, and phase of life. People's attitudes regarding whether to stay on or to leave depends also on the contextual circumstances in which the theme is explored. In dialogue with friends and acquaintances Aisha spends time producing prospects of how to establish a better life through people located elsewhere; how to get access to the resources necessary in order to do so. These discussions circle around spatial and social mobility, and in most discussions the two modes of mobility intersect. When 'opportunities for a better life' are on the agenda, the talk focuses on movement, not fixity; either in terms of moving towards a different life situation through marriage, to become part of a new household, or a temporary relocation to a different place. As discussed above, also Aisha has tended to emphasize her belonging in

13 Based on fieldwork in Tanzania he suggests that the imagination plays a predominant role in the cultural logics of migration. He has argued, 'While dreaming of migration is very important for young people's day-to-day life in Tanzania, travel abroad will not be a reality for most' (Salazar 2012, 18).

Zanzibar and thus her envisioned relocation to a different place would be talked about as temporarily and entail an eventual return home (*nyumbani*).

Apart from how Aisha's aspiration for social mobility reveal is shared by most Zanzibaris, her story is equally representative for all those who do not have access to resources that would make travel within the region, let alone international travel, possible. Moreover, once, when Aisha had a potential invitation for international travel, she was not granted a visa. Visa authorities at the European embassy in question concluded that, from their point of view, there was nothing – that is, business, family, wealth; not even her teen-aging son – that would secure her return to Zanzibar. Thus seen within such a framework, it appears as if it became impossible for Aisha to move, precisely because, according to the regulations followed by the visa authorities, she appeared not *settled* or properly fixed in place; she was without property or social relationships that were considered meaningful enough to oblige her to return. In this situation, her 'belonging' was understood with reference to particularistic economic and family configurations that were not immediately meaningful to her. Said differently, because, in their opinion, economically and emotionally Aisha would be afforded *to leave*, she would not be granted a visa. Interestingly however, the plans and choices she has made and still imagines, show that social and cultural 'boundaries' can be stretched and, even, transgressed. Yet, when confronted with economic and juridical perimeters, that is, the wider regime of mobility, at least, spatial movement is restrained.

Conclusion

In this chapter I have explored mobility and belonging through a focus on movement and trans-local and national connections as these are perceived and lived by people in Zanzibar Town – a society where most people have incorporated notions of multiple belonging and where the fact of having experienced life in other places, in itself, seems to provide a higher level of social standing. More specifically, I have explored how notions of mobility are inscribed in everyday-life practice and discourse. The two cases discussed suggest complex itineraries involving trans-local and national connections over time and generations and how in certain regions like the Indian Ocean, people's movements and interconnections have been constitutive of (new) societies. Approaching movement and trans-local and national inter-connections 'from below' it becomes clear that, over time, people are confronted with different forms of moral, cultural, and structural boundaries or what Glick Schiller and Salazar (2013) denoted as 'mobility regimes' (189). In the case of Zanzibar, the policies developed in the wake of the 1964 Revolution represented a deviation from what had, until then, been essential features of this society. From the 1960s, the newly independent nation-state and, especially, post-revolutionary Zanzibar aimed at limiting the forms of mobility that had been formative for the previous period, both within the western Indian Ocean as well as beyond, insisting on jurisdiction over territory (see Ho 2014). Despite the historical conditions of society, ideologically and identity-wise, for the post-revolution state the interconnectedness of the western Indian Ocean region was not associated with anything positive. On the contrary, the new state identified itself with continental Africa (Babu 1991, 244–5; Gilbert 2002, 32). Among other things, the limits placed, especially, on connections within the Indian Ocean after the 1960s meant that commercial and imperial structures that historically had produced coherence within the region faded and, in their wake previous routes and modes of connectivity were transformed or obstructed. In the period following the revolution, the main political idea was, precisely, to restrain social and economic life within a particular framework set by the government – including a control of all forms of connections in and out of the islands that were associated with the Zanzibari society prior to the revolution.[14] Since the late 1980s, the political ideology has again changed, turning tourism into the main

14 In 1986 the Zanzibar government established the tourism Investment Act, encouraging foreign investors to make proposals for hotel constructions. The Act resulted in a construction boom which reached its height during the period 1995–2000. By 2002, approximately 6,640 beds existed in 190 hotels and guesthouses (commission of Tourism 2002). Most of these were operated by the international tourist industry.

industry, attracting various foreigners to the islands – tourists, investors, and labour migrants.[15] Still, on the local level, people are concerned with what they see as political instability, severe economic stagnation, high unemployment, and a tourist industry owned and managed mostly by foreign people and companies, employing mainly immigrant workers (Gössling and Schulz 2005). People report that there has been a decline in public services, an increase in privatization of essential services in health and education, while also new and more market-oriented land laws and tenure systems have been introduced (see also Caplan 2009). In Zanzibar, currently it is mainly through connections across state boundaries, rather than within, by which people are associated with social mobility. Yet, when their aspirations imply a movement across state boundaries, projects usually stagnate.[16] Possibilities of travelling across national borders are perceived as increasingly difficult, at least for those who do not belong to the multi-local elites. My point here is that although openings and opportunities for spatial and social mobility are limited for most Zanzibaris, future prospects involving spatial movement are constantly nourished through local discourse. People belonging to different age-groups (*rika*) continue to engage in discussions about travelling, migration, and about places and life elsewhere. Women and men discuss potential migration and envision

15 The population of Zanzibar is about 1,211,000. Since the liberalization of the economy in the mid-1980s and the establishing of international tourism in the early 1990s, there has been a rather dramatic increase in population on the islands. In 1978 the population was 476,111 people, in 1988 the population was 640,685 and in 2002 the population had increased to 981,754 people ('Zanzibar statistical abstract' 2010, Zanzibar Official Statistics Provider; Socio-Economic Survey 2013).

16 Currently, in Tanzania in general most migratory mobilities are rural to urban, favouring, according to Noel Salazar, circular mobility as well as permanent migration to commercial attraction poles like the coastal cities of Dar es Salaam and other nearby East African countries (Prinz 2005, as cited in Salazar 2012, 14). As noted by Salazar (2010) Tanzania's National Bureau of Statistics provides no data on emigration either for Tanzania in general or for Zanzibar alone. However, they do provide statistics on, for instance, tourism. Zanzibar Official Statistics provider (2013) states that in 2012 about 170,000 visitors /tourists arrived in Zanzibar – which shows an increase from 2010 of about 133, 000 tourist arrivals.

better lives elsewhere and, eventually, upon an expected return, at home in Zanzibar.

This ethnography collected through a longitudinal fieldwork suggests that perceptions and practices of movement and interconnection are integral aspects of past and present livelihoods. Persons, families, and households have remained connected transnationally, and have continued to incorporate these connections and the various modes of mobility involved into the Zanzibari society. Cultural and moral aspects related to gender identity do, however, play a part when it comes to mobility. This does not imply that certain moral and cultural boundaries cannot be transgressed or that an organizational principle like sex segregation may necessarily restrict movability, but rather provide different modes of movement for women and men; while women move with husbands or in connection with marriages, men may move for adventurous and economic ventures.

Discussing mobility through the various narratives that people live and tell, I have tried to apply a perspective from 'below' to show how over time (im)mobility is lived and perceived and, moreover, how mobility is experienced in relation to culturally fashioned modalities of belonging and fixity. In terms of perceptions of movement and order, my main interest has been to explore social processes without pre-supposing neither sedentarism nor analytically speaking a strict territorialization of identity (Malkki 2002; Turton 1996). Spatial and social movements are more or less plausible for people differently positioned within society, precisely because of changing delimitations and boundaries caused by what Glick Schiller and Salazar (2013) denote as 'regimes of mobility' (189). Thus, emphasizing mobility while ethnographically focusing on processes of social and spatial motion, I have also found it pivotal to discuss to what extent local perceptions and practices related to movement and interconnections are responsive to changing 'mobility regimes'. At the local level and from the position of households, migration was and still certainly is motivated by a wish to secure a decent livelihood, but also because families, historically and memory-wise, may see themselves as inter-related with other places and life-worlds. An approach to the study of 'mobility' should, as discussed by Abdul Sheriff and Engseng Ho (2014, 19), privilege movements of people, the routes they follow, the relationships they mould, and the societies they create. In a long-term perspective the political, economic, and moral framework of movement and fixity, interconnection and rupture becomes modified and transformed and in its wake the life trajectories of people and persons' become moulded.

Bibliography

Allen, Tim, and David Turton. 1996. 'Introduction: In Search of Cool Ground'. in *In Search of Cool Ground: War, Flight & Homecoming in Northeast Africa*, edited by Tim Allen, 1–23. London: James Currey.

Alpers, Edward A. 2000. 'Recollecting Africa: Diasporic Memory in the Indian Ocean World'. *Africa Studies Review* (Special Issue on the Diaspora), 43 (1): 83–99.

Babu, Abdulrahman. 1991. 'The 1964 Revolution: Lumpen or Vanguard?' in *Zanzibar Under Colonial Rule*, edited by Abdul Sheriff. London: James Currey, 220–47.

Bang, Anne K. 2003. *Sufis and Scholars of the sea: The Sufi Family Networks of Ahmad Ibn Sumayt and the Tariqa Alawiyya in East Africa c. 1850–1925*. Curzon: Routledge.

Bascom, Johnathan. 1998. *Loosing Place: Refugee Populations and Rural Transformation in East Africa*. New York: Berghahn Books.

Bennett, Norman R. 1978. *A History of the Arab State of Zanzibar*. London: Methuen.

Bishara, Fahad. 2014. 'Paper Routes: Inscribing Islamic law Across the Nineteenth-century Western Indian Ocean'. *Law and History Review*, 32 (4): 797–820.

Black, Richard and Khalid Koser (eds). 1999. *The end of the Refugee Cycle?: Refugee Repatriation and Reconstruction (No. 4)*. New York and Oxford: Berghahn Books.

Campell, Gwyn. 1989. 'The East African Slave Trade, 1861–95: The Southern Complex'. *The International Journal of African Historical Studies*, 22 (1): 1–26.

——2005. *Abolition and its Aftermath in the Indian Ocean and Asia*. London: Routledge.

Cooper, Fredrick. 1980. *From Slaves to Squatters: Plantation Labour and Agriculture in Zanzibar and Coastal Kenya 1890–1925*. New Haven: Yale University Press.

Caplan, Pat. 2009. 'Understanding Modemity/ies: The Idea of a Moral Community on Mafia Island, Tanzania'. In *Knowledge, Renewal and Religion: Repositioning and Changing Ideological and Material Circumstances Among the Swahili on the East-African Coast*, edited by Kjersti Larsen, 213–36. Uppsala: The Nordic Africa Institute.

Eastman, Carol M. 1971. 'Who are the Waswahili'. *Africa: Journal of the International African Institute*, 41 (3): 228–36.
Fair, Laura. 2001. *Pastimes and Politics: Culture, Community, and Identity in Post-Abolition Urban Zanzibar, 1890–1945*. Ohio: Ohio University Press.
Ferguson, James. 1990. 'Mobile Workers, Modernist Narratives: A Critique of the historiography of Transition on the Zambian Copperbelt (part two)'. *Journal of Southern African Studies*, 16 (4): 603–21.
Freitag, Ulrike, and Achim von Oppen. 2010. 'Translocality: An Approach to Connection and Transfer in Area Studies'. In *Translocality: The Study of Globalising Processes From a Southern Perspective*, edited by Ulrike Freitag and Achim von Oppen, 1–2. Leiden: Brill.
Gilbert, Erik. 2002. 'Coastal East Africa and the Western Indian Ocean: Long-distance Trade, Empire, Migration, and Regional Unity, 1750–1970'. *The History Teacher*, 36 (1): 7–34.
Giles, Linda. 1987. 'Possession Cults on the Swahili Coast: A Re-examination of Theories of Marginality'. *Africa*, 57 (2): 234–57.
Glassman, Jonathon. 2011. *War of Words, war of Stones: Radical Thought and Violence in Colonial Zanzibar*. Bloomington and Indianapolis: Indiana University Press.
Glick Schiller, Nina, and Noel B. Salazar. 2013. 'Regimes of Mobility Across the Globe'. *Journal of Ethnic and Migration Studies*, 39 (2): 183–200.
Gössling, Stefan, and Ute Schulz. 2005. 'Tourism-related Migration in Zanzibar, Tanzania'. *Tourism Geographies*, 7 (1): 43–62.
Gurnah, Ahmed Salim. 1997. 'Elvis in Zanzibar'. In *Limits to Globalisation*, edited by Alain Scott. London: Routledge, 116–43.
——2008. 'Identity Through a Culture Haze'. Paper presented at The ZIORI Conference on the Indian Ocean, Zanzibar Town, Zanzibar, 15–17 August.
Harir, Sharif. 1994. 'Recycling the Past in the Sudan. An Overview of Political Decay'. In *Short-Cut to Decay: The Case of the Sudan*, edited by Sharif Harir and Terje Tvedt. Uppsala: Scandinavian Institute of African Studies.
Ho, Engseng, 2014. 'Afterword: Mobile Law and Thick Transregionalism'. *Law and History Review*, 32 (4): 883–9.
Horton, Mark. 1990. 'The Periplus and East Africa'. *Azania*, vol. 25, 95–9.
Horton, Mark, and John Middleton. 2000. *The Swahili*. Oxford: Blackwell.
Indra, Doreen Marie (ed.). 1999. *Engendering Forced Migration*. Oxford: Berghahn.
Kheshodkar, Akbar. 2010. 'Marriage as the Means to Preserve "Asian-ness": The Post-Revolutionary Experience of the Asians of Zanzibar'. *Journal of Asian and African Studies*, 45 (2): 226–40.
Larsen, Kjersti. 1990. 'Unyago – fra Jente til Kvinne'. *Oslo Occasional Papers in Social Anthropology*, vol. 22.

——1998a. 'Spirit Possession as Historical Narrative: The Production of Identity and Locality in Zanzibar Town'. In *Locality and Belonging*, edited by Nadia Lovell, 125–47. London: Routledge.

——1998b. 'Morality and the Rejection of Spirits: A Zanzibari Case'. In *Social Anthropology*, 6, 1. 61–75. Cambridge: Cambridge University Press.

——2000. 'The Other Side of Nature: Expanding Tourism, Changing Landscapes and Problems of Privacy in Urban Zanzibar'. In *Producing Nature and Poverty in Africa*, edited by Vigdis Broch-Due and Richard A. Scroeder, 198–220. Uppsala: Nordiska Afrikainstitutet.

——2004. 'Change, Continuity and Contestation: The Politics of Modern Identities in Zanzibar'. In *The Swahili and modernity*, edited by Pat Caplan and Farouk Topan, 121–44. Africa World Press.

——2008. *Where Humans and Spirits Meet: The Politics of Rituals and Identified Spirits in Zanzibar*. Oxford: Berghahn Books.

——2009. 'Introduction'. In *Knowledge, Renewal and Religion: Repositioning and Changing Ideological and Material Circumstances Among the Swahili on the East-African Coast*, edited by Kjersti Larsen, 11–38. Uppsala: The Nordic Africa Institute.

——forthcoming. 'Silenced Voices, Recaptured Memories: Historical Imprints Within a Zanzibari Life-world'. In *Memories of the Revolution*, edited by Marie-Aude Fouéré, Bill Bissell and Martin Walsh. Dar es Salaam: Mkuki na Nyota.

Lofchie, Michael. F. 1970. 'African Protest in a Racially plural Society' in *Protest and Power in Black Africa*, edited by Robert Rothberg and Ali Mazrui, 924–7. New York: Oxford University Press.

Makris, Gerasimos. 1996. 'Slavery, Possession and History: The Construction of the Self Among Slave Descendants in the Sudan'. *Africa*, 66 (2): 159–82.

Malkki, Liisa H. 2002. 'National Geographic: The Rooting of People and the Territorialization of National Identities Among Scholars and Refugees'. *Cultural Anthropology*, 7 (1): 24–44.

Middleton, John. 1992. *The World of the Swahili: An African Mercantile Civilization*. New Haven: Yale University Press.

Mitchell, James Clyde. 1959. *The Kalela Dance: Aspects of Social Relationships Among Urban Africans in Northern Rhodesia*. Manchester: Published on behalf of the Rhodes-Livingstone Institute by the Manchester University Press.

Myers, Garth, A. 2003. *Verandahs of Power: Colonialism and Space in Urban Africa*. Syracuse, NY: Syracuse University Press.

Pearson, Michael. 2003. *The Indian Ocean*. London: Routledge.

Prins, A. H. J. 1961. *The Swahili-speaking Peoples of Zanzibar and the East African Coast.* London: International African Institute (Ethnographic Survey of Africa Part XII).

Salazar, Noel B. 2010. 'Tanzanian Migration Imaginaries'. Working Papers, International Migration Institute, Oxford: University of Oxford, Paper 20.

——2012. 'The Power of Imaginaries in Transnational Mobilities'. *Identities: Global Studies in Culture and Power*, 18 (6): 576–98.

Sheriff, Abdul. 1987. *Slaves, Spices and Ivory in Zanzibar.* London: James Currey.

——1992. *The History and Conservation of Zanzibar Stone Town.* London: James Currey.

——2008. 'Maritime Culture and Globalization in the Indian Ocean'. Paper presented at ZIORI Conference on the Indian Ocean, Zanzibar Town, Zanzibar, 15–17 August.

——2010. *Dhow Cultures of the Indian Ocean: Cosmopolitanism, Commerce and Islam.* London: Hurst & Company.

——2014. 'Globalization with a difference: An overview'. In *The Indian Ocean: Oceanic Connections and the Creation of New Societies*, edited by Abdul Sheriff and Enseng Ho, 11–42. London: Hurst & Company.

Simpson, Edward, and Kai Kresse. 2008. 'Introduction: Cosmopolitanism Contested: Anthropology and History in the Western Indian Ocean'. In *Struggling With History: Islam and Cosmopolitanism in the Western Indian Ocean*, edited by Edward Simpson and, Kai Kresse. New York: Columbia University Press.

Tanzania Commission of Tourism. 2002. *Ministry of Natural Resources and Tourism*, Dar es Salaam, The United Republic of Tanzania.

Topan, Farouk. 1991. 'Reseaux Religieux Chez les Swahili'. In *Les Swahili entre Afrique et Arabie*, edited by Francoise Le Guennec-Coppens and Pat Caplan, 39–57. Paris: Karthala.

——2003a. 'Islam in African literature'. In *Encyclopedia of African literature*, edited by Simon Gikandi, 242–5. London: Routledge.

——2003b. 'Urban Folklore: The Swahili of Zanzibar'. In *African Folklore: An Encyclopedia*, edited by Philip M. Peek and Kwesi Yankah, 485–6. London: Routledge.

——2004. 'From Mwana Kupona to Mwavita: Representations of Female Status in Swahili Literature'. In *The Swahili and modernity*, edited by Pat Caplan and Farouk Topan, 121–44. Africa World Press.

Turton, David. 1996. 'Migrants and Refugees: A Mursi Case Study'. In *In Search of Cool Ground: War, Flight & Homecoming in Northeast Africa*, edited by Tim Allen, 96–111. London: James Currey.

Walker, Iain. 2010. *Becoming the Other, Being Oneself. Constructing Identities in a Connected World.* Newcastle upon Tyne: Cambridge Scholars.

Zanzibar Statistical Abstract. 2010. Office of Chief Government Statistician, Zanzibar Town: Zanzibar, June 2010.

Zanzibar Official Statistics Provider, Quick Figures 2012, *Socio-Economic Survey 2013, Statistical Report (Preliminary Results).* Office of the Chief Government Statistician, Zanzibar Town: Zanzibar, March 2013.

JENNIFER HAYS AND VELINA NINKOVA

4 Vertical and Horizontal Mobility among the Ju|'hoansi of Namibia

It is difficult to be far away from your home ground.
— JU|'HOAN man from the Omaheke

In this chapter, we examine patterns of social and geographic mobility among two populations of Namibian *Ju|'hoansi*.[1] The Ju|'hoansi are one group of *San*[2] peoples, former small-scale hunter-gatherers living throughout southern Africa. The San are often described in both academic and popular literature as 'nomadic' and 'egalitarian'. The first term implies a lifestyle characterized by high levels of *geographical mobility*, and the second implies a low level – or even an absence – of *social mobility*. In fact, anthropological and archaeological research has shown that the Ju|'hoansi and other San did have specific territories of resource utilization, within which they moved seasonally; they also depended on broad social networks to secure access to resources (Dieckmann, Thiem, Dirkx and Hays 2014; Lee 2013; Weissner 2002). Today, these patterns are changing as the

1 The plural *Ju|'hoansi* refers both to the people themselves and to the language. When referring to individuals, the correct term is *Ju|'hoan*; this is also an adjective. The vertical slash represents the click consonant made by clicking the tongue against the palatal ridge ('tsk'). Ju|'hoansi and other San languages contain many distinctive click consonants.

2 '*San*' is a term used to refer to a broader group of people who are former hunter-gatherers and who speak one of many languages characterized by click consonants. The Ju|'hoansi are one of these language groups, occupying central-eastern and northern-eastern Namibia, and central-western and northern-western Botswana. In this paper we are primarily describing the Namibian Ju|'hoansi, and we use the term 'San' when referring to the larger grouping.

San have largely been dispossessed of their land, and they are experiencing hierarchical social stratification as they are increasingly drawn into a cash economy. What can a 'mobilities' approach add to an analysis of the situation of former hunters and gatherers in southern Africa? What does 'mobility' – both geographical and social – mean to Ju|'hoansi communities and individuals today? How are these connected, and how has this changed over time? What is the role of education – a key element of social mobility and intimately linked with geographical mobility in a variety of ways? In this chapter, we apply the theoretical perspective of 'mobilities' to the case of the Ju|'hoansi, and propose a theoretical model for examining both geographical and social mobility and their intersections.

Mobility, socio-economic status and identity

In a recent description of the development of mobility studies, Glick Schiller and Salazar (2013) question whether *stasis* or *mobility* is the 'norm' from which vantage point we can address questions of physical mobility and socio-economic status. They describe new perspectives through which 'identities and loyalties could be understood as products of social relations rather than of fixed relationships to territory' (Glick Schiller and Salazar 2013, 186). While we agree with their general argument, we ask why should this be posed as an either/or question? Why should we understand identities and loyalties as products of social relations '*rather than*' as products of relationships to land? Glick Schiller and Salazar (2013) argue that 'movement and interconnection are fundamental to the human condition – past, present and future' (186). Their goal is to move away from binary thinking, to create a study of mobilities in which migration and stasis are seen as interconnected aspects of the human condition, and to do so in ways that provide a 'theoretical framework within which to address differential and multiple forms of power' (188).

In this chapter, we consider the Ju|'hoansi as a case study and propose a model for considering social and geographical mobility as interconnected variables. Our model is inspired by various aspects of mobility among Ju|'hoansi communities in the Omaheke region and Tsumkwe district of Namibia, but we are also interested in the extent to which it applies to – and how it differs from – other cases.

Both authors have extensive research experience with Ju|'hoan communities in Namibia and Botswana. Jennifer Hays has been working with the Nyae Nyae Ju|'hoansi, a small population in north-west Namibia, since 1998, focusing on education, livelihood, land issues, and indigenous rights. Velina Ninkova has conducted over a year fieldwork with a Ju|'hoansi community in another part of Namibia – the Omaheke – between 2008 and 2013. This chapter began as communications between us while Ninkova was in the field, about how an approach that places the concept of 'mobility' at the centre could help to understand the patterns that we have both recognized among the Ju|'hoansi. In the sections below, we briefly contextualize the Ju|'hoansi historically and geographically, describing in particular the two communities that we are discussing in this paper. We then propose a model that integrates social and physical mobility by placing them on two separate axes, and we explore how this model can be applied to the Ju|'hoansi. Before continuing with our main line of argument, however, we would like to briefly elaborate on the concept of *nomadism* (see also Engebrigtsen on nomadology in this volume).

Nomadism and the San

Almost universally, nomadism is described as an ecological adaptation in which people travel across territories with the purposeful intention to harvest or otherwise access resources seasonally or cyclically. Since nomadic movement entails the physical disassembling of the house or the household, nomads are 'mobile communities moving from one place to another without a fixed or permanent abode' (Gilbert 2014, 4). The anthropological literature describes three main groups of nomads: nomadic hunter-gatherers, pastoral nomads, and peripatetic service nomads (e.g. the

Roma in Europe as discussed by Engebrigtsen in this volume). Many, but not all, hunter-gatherers are nomadic or semi-nomadic, and belong to the first category (Gilbert 2014; Salzman 2002). Although there are commonalities that almost all nomadic groups share – in particular, conflict with local sedentarized populations – the differences between these groups are equally important for an understanding of mobility patterns. Most relevant here is that nomadic hunter-gatherers, while they may spend a large percentage of their time moving, can also have fixed territories, including permanent or semi-permanent settlements.

The San have often been presented in both academic and popular literature as a classic example of nomadic hunters and gatherers, moving frequently and lacking attachment to a fixed piece of land. Some San activists have publicly opposed this characterization, arguing that the term 'nomadic' both fails to capture the essence of their lifestyle and contributes to their ongoing dispossession and oppression.

> Our forefathers inherited the hunting grounds, water ponds, flora and fauna on Tc'amnqoo from their ancestors, and passed the ownership, control and utilization of this land down to us ... Migrations of Kua people would take place only within their territories. If a Bantu or a white man walks around his farm or field, no one calls him a 'nomadic person'. Yet we moved within our lands, within our marked tribal territories, and they called us nomads. This concept of 'nomadic' has been deliberately used to deprive us of the territories that duly belong to us. (Kiema 2010, 23–4)

> The people here speak of 'nomadism'. Why have they never asked us why we are going 'round. We are no nomads. There have always been territorial grounds that were respected by all. (San leader, as cited in Veber, Dahl, Wilson and Wæhle 1993, 149)

As the other chapters in this volume also show, a longitudinal perspective that puts patterns of mobility at the centre provides important insight into the dynamic relationship between 'mobility' and 'settlement how these patterns shift over time, and how they are connected to larger social and historical processes. The quotes above refer to cases in which the quality of 'nomadism' is interpreted as lacking fixed territory to have a right *to*, which is in turn directly linked to the forced sedentarization – or 'resettlement' – of San communities in Botswana and Namibia. However this settlement has in many places led to increased circulation when compared to traditional patterns of geographic mobility. Today, many Namibian Ju|'hoansi have little choice but to follow more distant and less predictable routes in pursuit of economic stability and subsistence (we will return to this point later in the chapter). With this in mind, we understand 'mobility' and 'stasis' not as opposite ends of a continuum, but as relative terms that are always interpreted within current economic, social, and political contexts (see also Chapter 1 in this volume).

From an anthropological standpoint, understanding the term 'nomadism' as a predominantly ecological adaptation fails to account for practices of recent former nomads who continue to conform to 'nomadic' values, norms, and practices after they have completely or partially sedentarized (Barnard 2002). Ingold (2000) has argued that people constitute their ideas of the world, not through passive transmission of information from one generation to another, but through active engagement with the environment. Toren (1993) suggested that a sense of a person's placement in the world 'is mediated by his or her engagement in the complex relations we describe under the rubric of "kinship," "religion," "political economy" and so on' (461). We see these theoretical perspectives as complementary, and recognize the importance of both the environment and social relations and institutions as two equally important aspects of the lived reality of people. In doing so, we break from the traditional view of 'nomadism' as an ecological adaptation only, and propose a more holistic understanding of the conditions behind peoples' mobility and interactions in space, time, and society. This argument is strengthened through our comparison of two related groups of Ju|'hoansi who exhibit very similar patterns of mobilities despite different historical processes and different levels of 'sedenterization'.

Background: Culture, land tenure, and social relations

> One of the factors that led to the large-scale land dispossession of the San during the pre-colonial and colonial eras was the erroneous belief that the San, due to their mobile lifestyle, never 'owned' land ... the San were perceived as nomadic ... as having no territorial claims.
>
> —DIECKMANN et al. (2014, 439)

The Ju|'hoansi were traditionally hunters and gatherers who lived in the north-eastern Kalahari, in what is now Namibia and Botswana.[3] Although colonial contact with other San groups in southern Africa began from the earliest years of European settlement, the Ju|'hoansi remained fairly isolated in the desert. In the Nyae Nyae area, there is no evidence of extended contact with other groups or integration into hierarchical systems before the mid-1950s (Biesele and Hitchcock 2011). In the Omaheke, intensive contact and settlement by white farmers, and the use of Ju|'hoansi as labourers, began very early in the twentieth century, and over the course of the next several decades steadily reduced their land base until virtually the only option for Ju|'hoansi was to live on the farms and work for the white farmers (Suzman 2000; Van Rooyen and Reiner 1995). Prior to colonization, both groups of Ju|'hoansi maintained a small scale, egalitarian social structure based on seasonal nomadism, with land-use patterns organized around territories called *n!oresi* (singular *n!ore*), composed of family-based units. The right to reside in and use the resources of these territories was inherited, and each *n!ore* was collectively owned by a small group of people, usually siblings, one of whom was identified as the *n!ore kxao*, sometimes translated as 'owner'. However, the role of the *n!ore kxao* (this was often a man, but could also be a woman) was primarily one of stewardship; of 'guardianship but not exclusive authority over the resources of the n!ore' (Biesele and Hitchcock 2011, 54). Permission to hunt and gather on another's *n!ore* had to be gained from the *n!ore kxao*; it was usually granted.

Resources were shared according to specific patterns, depending upon the nature of the item and upon social relationships (Lee 1993; Wiessner 2002). A particular type of gift exchange, *hxaro*, was characterized by vast networks based on the exchange of prestige objects. Wiessner (2002) says that 'Far-reaching networks of *hxaro* exchange allowed Ju|'hoansi to gain access to alternate residences when food, water, or social relations in their

3 Archaeological and linguistic evidence, historical records, rock engravings, oral histories, and more recently – genetic research, all indicate that the San were the first inhabitants of the southern African region, with archaeological evidence suggesting over 40,000 years of settlement in the region (Barnard 1992; Biesele 1993; Mitchell 2010, 2013; Smith 2001).

areas failed' (413).[4] These exchange systems were central to the survival of the Ju|'hoansi as a people and that of individual Ju|'hoansi. Biesele and Hitchcock (2011) call this 'a powerful, unwritten social technology of resource management' (54). This approach to resource management requires mobility, as groups and individuals circulate to access and distribute resources – and to maintain the associated social relationships. Importantly, these patterns of sharing and exchange occur over time, and are reinforced through specific name-sharing practices that span generations (see Ninkova 2017). Maintaining these social and economic relationships is an integral part of a long-term subsistence strategy; this point is also discussed in Chapters 5 and 6 of this volume.

Ju|'hoan culture continues to be characterized by a strong emphasis on egalitarianism and the maintenance of access to resources, and a high value is placed on exchange (rather than accumulation). A striking feature of Ju|'hoansi culture that persists despite numerous changes over the past decades is the active discouragement of self-promotion. Putting one's self above others, claiming to speak on behalf of others, or in some cases even performing well will earn one ridicule; one is expected to downplay one's achievements. Similarly the accumulation of material wealth generally results in increased demands to share (Hays 2002; Lee 1993). Biesele (1995) describes the dilemma faced by individuals who have access to jobs or political positions that provide them with differential access to resources:

> Ju|'hoan people, including those currently grappling with recent chances at unequal privilege, are only too aware of this strong feature of their society. They ask themselves, 'are special benefits worth losing status as an ordinary member of my society?' They weigh literacy, vehicle access and salary privileging against the familiar comforts of fair sharing and the security and 'anonymity' it conveys. (5)

We will return to the implications of this and related cultural values below.

4 Although in the Omaheke *hxaro* is no longer practised in the traditional sense, the word retains the meaning 'generous' (i.e. someone is *hxaro*).

Colonization and Namibian independence

These social systems based upon *n!ore* residence, and the associated patterns of mobility centred on the intersecting objectives of accessing resources and cultivating relationships shifted dramatically as both white and black farmers increasingly moved onto the traditional territory of the Ju|'hoansi, incorporating them into race-based hierarchical structures. The most explicitly racist of these was the apartheid system, which was implemented in Namibia, then South West Africa, under South African colonial rule.[5] During this period, 'homelands' were created for various linguistic groups in the area. However, as indicated in the quote above, the perception that San groups did not own land meant that homelands were not assigned to most of them – with the important exception of 'Bushmanland', then a remote desert area, part of which today is the Nyae Nyae Conservancy (described below).

After a long struggle by the South West Africa People's Organisation (SWAPO), independence was achieved in 1990. Land reform was a major concern at independence, and the new government immediately began developing strategies for redistribution at the National Conference on Land Reform and the Land Question held in 1991. In a series of legislative acts throughout the 1990s and early 2000s, various categories of land rights were delineated, and efforts were made to fairly redistribute land. The former homelands became communal lands, which allow for user rights to land – usually for farming or herding – on the basis of membership in a social group (defined by language, culture, and relational status). Other legislation provided for resource ownership – in particular, Wildlife Conservancies allow for control over the use of animal resources, and Community Forests allow for control over plant resources. In general the San have not benefited from this legislation, with the notable exception of the San living in the former homeland 'Bushmanland' (today Tsumkwe district) – this is described in the section on Nyae Nyae, below.

5 The country today called Namibia was a German colony – German South West Africa, between 1884 and 1915. Following the Second World War, it was taken over by the Union of South Africa, which extended race-based land allocation systems to its territory South West Africa, ultimately creating 'homelands' based on ethnicity.

Today, Ju|'hoansi outside of Tsumkwe district, including those in the Omaheke, live in one of several different situations, none of which provides land security. Farmworkers on commercial or communal farms owned by white or black farmers have no land ownership or usage rights at all, have no job security, and are highly vulnerable and mobile. Some Ju|'hoansi live in informal urban settlements, where they have no tenure security, and can be threatened with eviction. Some, particularly in the Omaheke, are living in Group Resettlement Farms under the Ministry of Lands and Resettlement. Although initially some support was provided to assist with agricultural development, these resettlement farms lack resources and economic opportunity, and the residents have no title to land (see Dieckmann et al. 2014 for a thorough description). In summary, low-impact land use strategies and small-scale egalitarian social structures have made it difficult for San throughout Namibia to resist the takeover of their land historically, with devastating results that continue into the present:

> The land used and owned by San groups in the past was summarily taken over, and this process continues unabated today as other ethnic groups come to occupy land which the San have been using traditionally for eons. The impacts of this land dispossession are manifold: economic dependency; loss of food security; social and cultural breakdown and alienation; inability to cultivate; withdrawal into marginal zones; high mobility; and, in some cases, inability to accumulate wealth in the form of livestock capital. (440, citing Suzman 2001)

Although there are common underlying patterns, these periods of colonization played out differently for Ju|'hoansi in the Omaheke and in Nyae Nyae, and have resulted in different situations today; we briefly outline some specifics of both sites below.

The Omaheke

Land dispossession in the Omaheke started in the beginning of the twentieth century, when waves of German and Afrikaner settlers entered the country (Suzman, 1999). Up until that point, the region was occupied primarily by San who had sporadic contact with scattered Bantu groups. One Afrikaner woman, the wife of a farmer, recalled to Ninkova that when their farm was first established in 1928, 'the Bushman were the only residents of the area but they were not settled and just wandered around' (field-notes 2008). The years that followed saw the incorporation of the Omaheke San into a racially based hierarchy in which they were at the very bottom, working as virtual slave labourers for white farmers (Suzman 2000; Sylvain 2001). Following Independence in 1990, many white settlers sold their land and left the country or dismissed their labourers, resulting in a wave of homeless and unemployed Ju|'hoansi.

Beginning in the early 1990s, the Namibian government began obtaining farms and resettling Ju|'hoansi, providing them with basic services and start-up funds for small subsistence and income-generation projects. However, access to markets in which they can sell items produced has remained very limited and inconsistent. Furthermore, the land does not belong to the Ju|'hoansi, and they have limited access to traditional resources. Farming activities, such as livestock breeding and crop production, are defined as the only legitimate land utilization activities; hunting is not permitted on the resettlement farms. The accumulation of people on small parcels of land results in the fast depletion of bush foods, hindering gathering. Without any secure access to land, the majority of the Omaheke Ju|'hoansi are today forced to seek employment as seasonal or permanent labourers on commercial or communal farms or support themselves through small jobs ('piece work'), subsistence farming, and government aid on the government-owned resettlement farms or in the squatter camps around the main town of Gobabis (see also Dieckman et al. 2014).

Nyae Nyae

Further north, in Nyae Nyae, the Ju|'hoansi had fairly little contact with either white or black groups until the mid-twentieth century. In 1959, the colonial South African government established the central town of

Tsumkwe with the intention of 'sedentarizing' the Ju|'hoansi, who were believed to be nomadic. The Odendaal Commission was set up in 1964 to produce a plan for separate development in Namibia, based on the extension of apartheid throughout the territory. 'Bushmanland' was designated as a homeland for San in Namibia in 1976. This area corresponds with the Tsumkwe district today, the eastern part of which is today the Nyae Nyae Conservancy (see below). The establishment of homelands had devastating effects on the San throughout the country. Although the Ju|'hoansi of Nyae Nyae did receive a 'homeland' of their own it was only a fraction of their ancestral territories. San living outside the reserve suddenly found themselves incorporated into the ethnic homelands of other groups, even if the land they had been previously living on was their own ancestral land (Gordon and Sholto-Douglas 2000, 173, 222).

As their other options diminished with the encroachment of both white and black groups onto their land, people began to move to the town, and in the 1970s the South African army established a base there. By the 1980s, living conditions in Tsumkwe had deteriorated dramatically, people were highly dependent and diet and health were poor. Most Ju|'hoansi expressed a desire to return to a social and resource management system based on the *n!ore*. With assistance from a development organization, the people moved back out to re-establish their *n!oresi* (see Biesele and Hitchcock 2011).

In 1998 Nyae Nyae was given the status of a Nature Conservancy, with resource rights based on ethnicity-linked membership. Specifically, Conservancy status allows members control over wildlife resources. In 2013 the Nyae Nyae Conservancy was gazetted as a Community Forest, providing them with rights over plant resources as well. Thus, today the Ju|'hoansi of Nyae Nyae have the right to hunt and gather on their own territory, and to benefit from their resources through tourism, professional hunting, and sale of products they gather or produce. In comparison to other Ju|'hoansi and other San groups, the Nyae Nyae Ju|'hoansi are very fortunate to have access to land and the right to manage their own resources. However, despite the legal protection that the Conservancy and Community Forest are meant to offer, their territory and resources are still under threat from neighbouring groups who wish to graze their cattle there. In Nyae Nyae today, there is a high level of mobility from village to village, and especially between the villages and the town of Tsumkwe.

Mobility axes

To better understand the kinds of mobility that the Ju|'hoansi engage in, we propose a model for looking at mobilities as an intersection of social and geographical mobility on horizontal and vertical axes (see Table 1). We define these types of mobilities as follows:

- *Socio-economic mobility:* social movement, within or among status levels (this can include moving up and down a social ladder, or networking among equals).
- *Geographical mobility:* movement in space for any of numerous reasons.
- *Horizontal:* movement that does not necessarily change socio-economic status.
- *Vertical:* movement that has the potential to significantly change status.[6]

We are using the term *socio-economic* rather than simply *social* mobility in order to emphasize the link between social status and economic opportunities. However, below we challenge some of the assumptions about the nature of these connections. In the remainder of this chapter, we shall explore each of these intersecting mobilities as they relate to the Ju|'hoansi.

6 'Vertical nomadism' is sometimes used to refer to the practice among nomadic pastoralists of moving animals to higher or lower elevations, following seasonal weather changes. In our model, we refer to vertical *mobility*, which does not refer to changes in physical elevation, but rather refers to the intersection between socio-economic status and travel possibilities.

Table 1

	Socio-economic	*Geographical*
Horizontal	*Internal*: Maintenance of as many connections as possible within ones social group; maintaining a network or 'safety net'	Physical movement that does not change socio-economic status; usually travelling short distances requiring little economic input
Vertical	*External*: 'climbing the social ladder'; gaining or losing wealth and status within a hierarchical social system	Physical movement that has the potential to significantly change status; usually requires substantial economic input

Socio-economic mobility among the Ju|'hoansi

The historic dispossession of their traditional territories that the Ju|'hoansi have suffered over the last century has led to the loss of a significant number of traditional practices intimately related to the land and its resources. Despite these forces, many San groups have shown enormous cultural resilience in the face of change, and have preserved many aspects of their traditional foraging ethos, most significantly their extensive kinship networks, as well as their strong egalitarian ideals (see for instance Guenther 1999). Therefore, we are careful to distinguish between 'internal socio-economic mobility' within one's own society, and 'external socio-economic mobility' that includes movement between the various sectors of society; for the Ju|'hoansi such movement generally entails leaving one's own group. As described above, among the Ju|'hoansi, there is a general lack of aspiration to visibly progress 'upwards' within their communities. There is a distinct avoidance of putting one's self in a position that is 'above' others within the community or in which one has 'more' of anything – including status, recognition, money, food, or other goods. This egalitarian ethos has been thoroughly described by numerous researchers, for the Ju|'hoansi in particular and as a general characteristic of San culture (e.g. Biesele 1993;

Guenther 1999; Hays 2016; Lee 1993; Wiessner, 2002). Although this is changing somewhat, and although today some Ju|'hoansi have jobs, earn money, and enter into political office (we will discuss this below), in general, the ethos of egalitarianism still strongly defines social relationships.

Horizontal social mobility

The maintenance of good social relations within Ju|'hoan society is very important; this is what we are identifying as *horizontal* socio-economic mobility. This type of mobility requires an active role on the part of each individual – social networks need to be lubricated on a regular basis and individuals will prioritize the maintenance of these networks above many other activities. Efforts towards this goal include regular visits between relatives and friends (nearby or far away), complex systems of sharing and exchange, and the observance of respect among people in the network. *This 'networking' is integral to their social, economic, and physical well-being individually and collectively.*

Because the networks they are maintaining extend geographically across a physical landscape, people are required to be physically mobile in order to maintain them and will often prioritize these movements. For example, in the Omahake, a Ju|'hoan family consisting of four generations and living in one of the resettlement farms used the little income they had to keep their horse and horse cart in good shape. They used this transportation weekly for shorter trips to neighbouring settlements to run errands or meet with people. At least once a month, they would also make a longer trip of up to several days to a commercial farm where close family lived. These longer trips are often accompanied by small gifts, food (meat being the most valued) and news being brought home to the resettlement farm.

People without access to vehicles walk far distances for errands that also give them the opportunity to meet with friends and relatives. People sometimes expressed appreciation for the act of walking itself; a young man in the Omaheke who would often walk 20 or more kilometres for a small box of snuff tobacco, sometimes in the dark early morning hours told Ninkova 'when you sleep poorly, it is good to walk and freshen up your mind' (Ninkova, field-notes 2013). Cell phone communication has also flourished, and kin living far from each other use this medium to exchange invitations to visit. Not honouring an invitation is considered extremely rude, and people often pack their things and move for extended visits lasting from a few days to a few months (Ninkova, field-notes 2013).

Similar patterns exist in Nyae Nyae and circulation among the villages is common, particularly with villages nearby but sometimes also across the conservancy; often these visits are primarily for the purpose of 'visiting' and securing social networks. Even more regular is movement back and forth between the villages and Tsumkwe. People travel to the town to gain access to food or other resources (in particular, drought relief food and old-age pensions, which are distributed in the town), to go to the clinic or to access other services available in Tsumkwe, or to look for work. These visits are almost always also social, and sometimes that is the main purpose of the travel even if other reasons are given for the trip. As in the Omaheke, people articulate their willingness to travel great distances in search of tobacco, and occasionally contrast this with alcohol; one man said, 'we will travel *far* for tobacco, but alcohol, we only drink if it is there in front of us' (Hays, field-notes 2001).[7] In general, while smoking is considered social, alcohol is recognized to damage social relationships

The vast majority of Ju|'hoansi, in both the Omaheke and Nyae Nyae, do not have a motorized vehicle at their disposal for the travel necessary to maintain these networks, thus these journeys are mostly by foot, sometimes by horseback or donkey. When possible, people try to get a ride with someone who has a vehicle (we will return to this in the section on geographical mobility, below). But in almost all circumstances, the initiative and the input required (including transportation) to maintain these networks is up to the individual.

7 This does not mean that people never travel for alcohol; surely some do. But it is far more common for tobacco to be the expressed reason for a trip to another village or the town.

'Vertical' social mobility

The other kind of social mobility is *vertical*, which is what we generally think of as socio-economic mobility. Vertical mobility 'upwards' is generally the aim of development projects, training courses, and education in general. 'Progress' is primarily seen as enabling individuals to get a job and earn an income. Namibia's aspiration to cut the social and wealth gap in the society through becoming 'a learning nation' (Government of the Republic of Namibia 1999) reflects the widespread international trend linking social mobility and education, as two necessary and desirable components of human progress. Furthermore, the jobs associated with upwards social mobility are usually available primarily in urban areas; likewise travel to – and residence in – urban centres is intimately linked to upward mobility.

The website of the Namibian government's *San Development Programme*, under the Office of the Prime Minister, says on its home page, 'The main objective is to integrate marginalised communities into the mainstream of our economy and improve their livelihood' (par. 4). Although the language refers to communities, in reality integration into the mainstream almost always occurs at the individual level, with benefits then spreading to relatives. This is explicitly expressed by most San youth when they talk about their aspirations regarding school; ultimately many say they aim to 'get a job and help my family'. While many San individuals express a desire for such integration – through successful employment – many do not want to have to choose between these opportunities and their current social networks. If we understand these social networks as also playing an important economic role for individuals and the general community, we can also understand the choice to physically leave the community as one that involves substantial risk. There is a general assumption underlying development efforts that are characterized by an emphasis on jobs and income that people want to 'escape' from their immediate circumstances of poverty, and that such escape involves leaving the physical place that they are from. We would like to problematize this assumption, and ask whether people do in fact want to escape, and if so, in what ways? In exchange for what, and at what risk?

Since we are using the metaphors of horizontality and verticality to talk about social movement, we would like to point out the role of *gravity* in socio-economic mobility. Unlike horizontal progression, in which one is travelling along a level field, in vertical progression one is always battling forces that will pull one back down. One cannot just move upwards without any support – there has to be some kind of scaffolding – a 'ladder' to climb. Where does this support come from? In the following section, we illustrate some of the complexities involved in 'climbing the social ladder' by examining support for schooling.

Schooling and social mobility

The literature on the correlation between education and social mobility suggests that human capital alone is not enough for successful progression through the education system. Of equal importance is the 'social capital' the family or the community possesses, including what value it places, and how much time it spends with, their children (Coleman 1988). While this presupposition that human action is driven – both individually and collectively – by desire for external social mobility may hold true for many societies, especially in the western industrialized world, there is a growing body of research suggesting that the relationship is not straightforward and unambiguous (see Froerer 2012 for an example from India; see Portisch 2012 for examples from Mongolia). Our data also support these findings, in particular in that parents, students, and the community in general prefer to invest their social capital in the maintenance of internal horizontal networks more than they do on external vertical mobility. This is often interpreted by government officials and other outsiders to mean that the Ju|'hoansi and other San 'do not care about education' or need to be convinced about the 'value of education' – both standard phrases in discussions about the low participation rate of San children in formal education (Hays 2007, 2016).

According to our interviews and observations over the years, however, there is explicit recognition among the Ju|'hoansi of the advantages that education can provide, and if they are able, the family does try to provide support. In recent fieldwork in Nyae Nyae, Hays found that the factor that most strongly predicted a student's likelihood to continue at the government school past Grade 4 was whether or not they had a family member who was employed in Tsumkwe, the town where the mainstream government school was located (Hays, Hopson and le Roux 2010; field-notes 2014). This support was critical to determining whether or not they stayed in school – but only up to a certain point. Several related points must be emphasized here. First, this illustrates how horizontal mobility – the maintenance of local social relationships – can play an important role in vertical social progression – these different 'planes' of movement are not separate but connected in complex ways.

In addition, it must be made clear that such support is by no means a *guarantee* of success in school. The Ju|'hoansi, and other San, face enormous challenges in the formal education system, of which a general lack of economic support is an important one. Although San students in Nyae Nyae and in the Omaheke are exempt from school fees, there are many other material needs, including food, clothing, shoes, hygiene items, school supplies, and lodging; lack of any one of these can be enough to prevent a child from continuing with school. Barriers related to language, culture, stigma, and other dynamics can cause children to drop out even in those instances when basic material needs are met. In short, even among those Ju|'hoansi who continue for a few years at Tsumkwe School, the vast majority drop out before the end of primary school. In the Omaheke, Ju|'hoansi who attend secondary school at boarding schools in Gobabis or elsewhere also drop out at very high rates (this is less true for those who attend secondary schools closer to their own communities; but this is an option only for some).[8] Since participation in school often entails both economic investment and social sacrifice (because children must live far away from their parents and other relatives), it is a risky investment (see Dieckmann et al. 2014; Hays, Hopson and le Roux 2010).

There are few Ju|'hoansi who earn enough money to pay for school fees and material necessities for their children (or those of relatives), thus

8 There is a secondary school in Drimiopsis and one in Post 10.

most Ju|'hoan children drop out of school once material support becomes necessary – unless support is provided by outsiders. 'Outsiders' can be the state, non-government organizations (NGOs), churches, or benevolent individuals (from dominant Namibian groups or expatriates), who provide economic and other support for students to encourage them to stay in school. Sometimes such support takes the form of ad hoc gifts of money, uniforms, or supplies – while useful, such help is neither consistent nor predictable. In order for San learners to receive regular support from government or non-government sources, they have to meet certain criteria – in terms of grades, attendance, or behaviour – which they are not always able to. Accessing external support may itself require knowledge or resources that people do not have (transportation, money and materials for postage of applications; familiarity with the language of forms they must fill out, and so on). For these and other reasons, despite the availability of support, there are very few Ju|'hoansi (or San in general) who have achieved high enough levels of education to allow them to obtain positions where they receive a steady income, or who are in positions of political authority (and those who do face other challenges; see the following section).

We would like to also emphasize that receiving a degree or qualification does not automatically translate into securing a job. The reverse is also true – formal education is not necessarily required in order to have a job or an income. Among the relatively few Ju|'hoan individuals who do have secure jobs or who are in positions of local political authority, many have not received a degree, but hold these positions because of personal characteristics or skills. These include, for example, traditional authorities, translators, drivers, assistants in development projects and tourist guides, among others.

Risks linked to vertical social mobility

Finally, and importantly, those Ju|'hoansi who do achieve 'success' in terms of greater access to resources and greater geographical mobility become vulnerable in other ways. If they remain in their communities, they often experience serious tensions with their relatives. Individuals with jobs report that they are responsible for many others – they can expect both to be depended on by others *and* to be the target of intense criticism and ridicule. Such techniques used to maintain social equality are often referred to in the anthropological literature as 'levelling mechanisms' (Lee 1969; see also Willerslev 2009, 81–2 for a description of how this works among Siberian Yukaghirs). These mechanisms, which can be unpleasant for the person who is the target, serve both to ensure that people who gain accesses to resources will distribute them, and also that no one person will gain prestige or power over others. All individuals (or families) who gain vertical mobility will be expected to assist the people who are in their 'network' unless they decide to remove themselves from this network.

Examples of people in such positions include (among others) teachers, people working for NGOs, and people in positions of political leadership, either within the Namibian government structures or within the Traditional Authority system.[9] The two traditional San Chiefs in the Omaheke (one of whom is Ju|'hoansi; the other one is !Xoon) – both face difficulties negotiating their positions of power and visible wealth (vehicles, cattle, monthly salary) among their communities. They are the subject of frequent gossip even among their extended families, and are often criticized for using their position to provide financial or other benefits to their own families. In Nyae Nyae, a common complaint is that the chief does not come out to the villages enough. Although the Traditional Authority is supposed to be provided with a vehicle by the government, these vehicles are often not working, limiting the ability of the Chiefs to consult with their constituency. It is important to point out, however, that criticism of the chiefs does not necessarily mean they are not legitimate, or that all of the critiques should be taken at face value. The levelling mechanisms described

9 The *Traditional Authorities Act* (2000), under the Ministry of Regional and Local Government and Housing, allows a community to designate one person to be their chief, in accordance with customary law, who then has to be approved by the minister. The 'community' is defined in the act as 'an entity sharing a common language, culture and heritage'. There are two San chiefs in the Omaheke, including one who is Ju|'hoansi; there is also a Ju|'hoansi chief in Nyae Nyae (see Dieckmann et al. 2014).

earlier apply even – perhaps especially – to those whose positions are highly valued by the community. Both Ju|'hoansi chiefs are considered by their communities to be legitimate leaders who were selected through legitimate processes. A similar situation exists for teachers, in particular in Nyae Nyae where the village schools employ several Ju|'hoansi as teachers in mother-tongue lower-primary schools located in the villages (see Hays 2002, 2016). It can at times be difficult for outsiders to differentiate between serious community dissatisfaction (with political representatives, teachers, or other employees) and complaints based on resistance to social stratification.

Not all people who are 'upwardly mobile' do so in particular service to the community. A Ju|'hoan man from a resettlement farm in the Omaheke who is wealthier than the majority of Ju|hoansi in the settlement, and who plays an active role in community decision-making and is vocal at communal meetings, is facing increased backlash from the community. He is referred to as 'Xhosa' or 'Zulu',[10] indicating that he is assertive and greedy, which is often associated with these ethnic groups. Young men would sometimes go to great lengths describing how they would physically harm him, often in an exaggerated manner. In addition, details of his personal life surface often, serving as a reminder that he is not in a position to judge and order others around (Ninkova 2017).

Some 'upwardly mobile' Ju|'hoansi partially escape this dynamic by moving away from their communities; others spend time away from their home because it is required by their training or their job. Being separated from their communities and territories is usually expressed as a great hardship (see also Chapter 5 in this volume). A young San woman who worked in a police station in northern Omaheke expressed her frustration with the job, with the distance she was forced to live apart from her family in the south of the region, and with the pressure she was feeling because of this. She was married, had a young child, and said that she wanted to leave the job and go back to her family in the south. Most Ju|'hoansi youth in the Nyae Nyae area express a desire to remain close to their home village, or at least in the Conservancy. Among potential employers, the San have a

10 These are the names of Bantu language groups, most of whom live in South Africa and whom this community perceives as being particularly arrogant and assertive.

reputation for leaving jobs and are thus often considered 'unreliable' – when asked, San individuals will often say they missed their family and village and decided to return home. 'It is difficult to be far away from your home ground', a young Ju|'hoan man exclaimed to Ninkova, referring to the hardships one is presented with when removed from their support network and illustrating the link between place and social belonging.

Can one simultaneously maintain horizontal networks and pursue vertical pathways?

According to our observations, it seems difficult for Ju|'hoan individuals to be vertically mobile without 'distancing' themselves to a certain degree – either physically or socially – from the social networks upon which they depend for survival. Nonetheless, some people do try to do this, and some individuals are able to find a way to balance the competing demands and contradictory value systems. Often, this effort is linked with geographical mobility, which entails both physical separation from their group, and also the potential for greater opportunity through access to training or jobs. There are risks in doing this, which will be addressed in the next section.

Geographical mobility: Village/Town/International

Most Ju|'hoansi in Nyae Nyae and in the Omaheke live in small remote settlements or on farms (in the Omaheke) with few services or resources immediately available to them. Larger villages may have small shops, but in order to access basic food items, pension checks, and other goods and services, or sell one's produce or handicrafts, one must travel between the villages and the larger central towns. For the Nyae Nyae region this town is Tsumkwe (population of approximately 2,500), which is in the centre of the Conservancy, in which all villages are inhabited by Ju|'hoansi. In the Omaheke, the central town is Gobabis (population approximately 20,000), which is much larger than Tsumkwe and which is surrounded by farms, including those where Ju|'hoansi and other San have been resettled. While most of the jobs in both of these towns are held by non-Ju|'hoansi, a few Ju|'hoansi are employed and live in houses there. In both towns, squatter camps have also developed around the edges and many Ju|'hoansi reside in these locations. In both areas, geographical mobility among the Ju|'hoansi takes place at different scales, which we briefly elaborate on here.

Horizontal geographic mobility

Horizontal geographic mobility is a metaphor for circular movement between settlements, farms, and town, for the sake of subsistence or conflict avoidance, for maintaining social networks and accessing services not available in one's own settlement, among other reasons. This movement is characteristic for all Ju|'hoansi in Nyae Nyae and the Omaheke, and is the most secure form of long-term insurance in times of decreased resources or hunger. The horizontal path an individual follows can extend to hundreds of kilometres away from one's home. An old man living in Skoonheid in the Omaheke, for example, has children and other family members living on farms and communal areas stretching from Otjinene in the north, Epukiro Post 3 in the east, on a farm close to Windhoek in the west, together with numerous farms close to his home, and makes visits to all of these on a regular basis.

During the past few decades the concept of 'town' and the towns themselves, have begun to play an important, but ambiguous role in Ju|'hoansi discourse. It has in a way become an intersection of the two types of mobility they can explore. On one hand, 'town' is a very important part of the horizontal mobility network of many; Ju|'hoansi living in both Tsumkwe and in Epako (the informal settlement of Gobabis) maintain active networks with their relatives in surrounding farms and settlements.

Town has also become a symbol of vertical social movement – importantly, education opportunities past primary and most education-related jobs are to be found there – and many view it as a symbol of progress and opportunity.

Ju|'hoansi in both Nyae Nyae and the Omaheke conceptualize some aspects of 'town' as positive. For example, in town one can get a job, or participate in other social exchanges – and being excluded from either of those opportunities is perceived as problematic. One young woman from the Omaheke said that the government had resettled them far away from Gobabis, so that they would be cut off from information about what is happening in the town and would remain ignorant. At the same time, the town itself is associated with things that make people feel uneasy – alcohol, a fast pace, noise, and crime. Not knowing anyone, or not having family where you can spend the night or eat is a treacherous prospect for many. After visiting town most prefer to go back to their settlement or village as soon as possible. A young man from the Omaheke once reported, 'In town, you feel like a mad man'. Tsumkwe was once known as 'the place of death' and is still sometimes referred to in similarly negative terms.

Vertical geographical mobility: Windhoek

While travel between villages or between the village and nearby towns is a fairly common occurrence for many Ju|'hoansi, travel to larger urban centres like the capital city, Windhoek is another matter. In order to get to the city, one needs access to a vehicle, or to the money to pay for a taxi, a bus, or a lift in a private car. Such trips thus require financial resources that most Ju|'hoansi do not have, and the vast majority of such trips are made with funding and/or transportation from an outside source. This could be, for example, the government paying for or arranging the transportation of members of the recognized Traditional Authorities (to attend a meeting), or for specific artisans or community representatives to travel to a workshop or other meetings with support from the government or non-government organizations.

Some San, including Ju|'hoansi, live in Windhoek for extended periods for training courses or higher education. These individuals are supported through scholarships and stipends, given either by government or NGOs. A recent nationwide survey of San in Namibia conducted interviews with several San living in Windhoek either as students or as government employees, and found the following:

> San interviewees in Windhoek considered their living conditions in this city to be better than in their home communities. This was at least partly due to the fact that they were … benefiting from a regular income. At the same time, this income was necessary for them to be able to live in the city. If there were problems and the income ceased or became irregular, it was far more difficult for them to get by than it was in their home communities. Some of the interviewees were residing with immediate family members in Windhoek, and others were alone, but all of them lacked the extensive social support networks that would have supported them in difficult circumstances back home. So, while appreciating the advantages of living in Windhoek, some interviewees expressed a preference for being closer to their families and support networks. It was clear that an essential precondition for living in Windhoek was a dependable source of money. (Dieckmann et al. 2014, 538)

Such positions are risky for Ju|'hoansi who are vulnerable to the decision-making processes of others who are not from their own group. Their funding or employment is subject to delay, interruption, or termination. Without the safety nets of their horizontal networks, if the support on which they depend to survive in the city is taken away, they could plunge quickly.

International travel

International travel is of primarily two types. Ju|'hoan communities straddle the Namibia/Botswana border and individuals occasionally travel across; most often such trips are to visit relatives. Although these visits are socially horizontal in purpose – visiting family and cultivating exchange relationships on the other side of the border – such travel requires travel documents and some resources for transportation, which are limiting factors for many Ju|'hoansi. But those who have obtained travel documents enjoy moving back and forth across the border to exchange information and gifts with relatives.

Some Ju|'hoansi have travelled by plane to faraway places such as Norway, Australia, the USA, and Great Britain, often to participate in conferences relating to indigenous peoples or for meetings. Such travel is virtually always funded by donors, including private individuals, NGOs, and sometimes the government.

Mobilities square

We now return to the mobilities square presented earlier. Given the complex relationships between social mobility and geographical mobility, we have found it useful to imagine the horizontal and vertical mobility of the Ju|'hoansi on this axis (see Table 2).

Table 2

	Socio-economic	*Geographical*
Horizontal	*Internal*: Maintenance of as many connections as possible network; 'safety net' • Elaborate naming systems • Hxaro, exchange • Obligation to share, according to object and/or relationships	Movement that does not change socio-economic status; usually travelling short distances requiring little economic input • Seasonal movements for food procurement, or employment • Travel for exchange, relationship maintenance • Movement for conflict resolution • Movement for other reasons *Means: foot; donkey/horse; lifts*

	Socio-economic	*Geographical*
Vertical	*External*: 'climbing the social ladder'; gaining status and increased wealth among others • Requires a job, income • Higher education levels are associated with higher salary, and higher mobility • 'levelling mechanisms' of Ju|'hoansi strongly discourage 'rising above' others • Compatible with remaining in the network?	Movement that has the potential to significantly change status; usually requires substantial economic input • Movement to urban centres, usually for jobs or education • Travel for training opportunities • Travel to meetings in urban centres, international travel *Means: lift, car, bus, airplane*

While vertical mobility is often associated with 'empowerment', it is worth noting that in the Ju|hoansi case, empowerment comes mostly from maintaining a wide horizontal movement. Horizontal movement is circular in nature, based on reciprocity and back and forth movement in which an individual plays an active role in the engineering of the movement itself (see also Chapter 5 in this volume). If a relationship is damaged, there's always a chance for repair (Ju|'hoansi in the Omaheke and Nyae Nyae consider a conflict resolved when the 'guilty' party apologizes and the other accepts the apology). If food resources are depleted, there is always a new territory, or family members, to be explored. Each individual bears the consequences for his or her actions.

On the vertical axis, however, the San are dependent on others to provide them with an opportunity to prove themselves, and if they fail, they seldom get the chance to repair their wrongs. This situation is very much resisted by many Ju|'hoansi who place an extremely high value on the personal autonomy of the individual as a sole decision-maker in his or her actions and fate.

Conclusion

For the Ju|'hoansi, we can ask whether *horizontal* and *vertical* mobility are compatible or whether they require modes of operation (including values, behaviour, physical mobility, economic activity, personal comportment, ways of relating to others in their networks) that are difficult to reconcile. How does this play out in other groups? We are interested in whether others find that this model applies to situations that they are familiar with, and if the questions we are asking are relevant. Can one simultaneously maintain both horizontal networks and pursue vertical pathways? Can we think of *horizontal mobility* in both geographical and social terms as the development of a network of equals, as opposed to a vertical 'ladder' that one climbs – and can this be recognized and valued by government and donor-driven development efforts? How can we value and recognize horizontal mobility and simultaneously acknowledge the desire of individuals and groups to gain 'capital' (social or economic)?

Looking at situations like that of the Ju|'hoansi and other small-scale and egalitarian groups through the lens of mobility provides an important perspective. Everything is always in motion – there is no such thing as stasis, only different tempos and patterns of movement. Describing the ways in which people move, for what reasons, and the power relations involved can also allow us to better understand why, for example, the Ju|'hoansi (and other San) make choices that often seem counter-intuitive from a perspective that privileges vertical socio-economic mobility, and 'vertical geographic mobility' (as we have described it in this paper). Governments, donors, and international or local organizations often promote a 'vertical' approach to development – aiming for a measureable increase in standard of living, or numbers of children in school.

Vertical mobility – social and economic – is also desired by the Ju|'hoansi; it is expressed both by some individuals (often with the ultimate goal of helping their families), and as a general desire of the group to have some of its members gain stability within the mainstream economy. However, as we have described in this chapter, such mobility can entail a large social and economic risk. It often also entails an enormous sacrifice of personal autonomy, as one becomes bound by requirements and circumstances that are out of one's control, as well as by the demands of kinfolk for financial assistance. Many Ju|'hoansi consciously choose *not* to follow this vertical path, and instead continue to cultivate their horizontal relationships – in large part through the horizontal geographic mobility as described in this chapter.

Neither the vertical nor the horizontal axes completely explain Ju|'hoan practices of either socio-economic or geographical mobility; these variables intersect in complex ways. Furthermore, the dynamics we have described above are also unfolding over time. A longitudinal perspective provides insight into which cultural strategies are likely to be adaptations to a particular situation, and which are maintained over time. The dynamics that we have discussed in this chapter reflect responses to current economic and social situations – but which are clearly based on long-term resource management strategies.

As noted above, an assumption of idealized vertical progression underlies much development work today. Within that perspective, both the mobility patterns and resource distribution strategies of the Ju|'hoansi are often interpreted as short-term solutions of people who are 'living in the moment' (Suzman 2000, 48), and these characteristics are said to be reflected in an inability of San to make long-term decisions (Hays 2016). We suggest, however, that these strategies of the Ju|'hoansi are in fact integral to a long-term strategy that is based upon the cultivation of horizontal social networks across space and time. We wish to emphasize that only by recognizing this horizontal axe – and its legitimacy and efficacy as a long-term risk-management strategy – can any approach aiming to increase food security and overall quality of life for the Ju|'hoansi and other San be effective.

Bibliography

Barnard, Alan. 2002. 'The F Mode of Thought'. In *Self- and Other-Images of Hunter Gatherers. Senri Ethnological Studies no. 60*, edited by Henry Stewart, Alan Barnard, Keiichi Omura, 5–24. Osaka: National Museum of Ethnology.

Biesele, Megan. 1993. *Women Like Meat: The Folklore and Foraging Ideology of the Kalahari Ju|'hoan.* Bloomington: Indiana University Press.

——1995. 'Governmental and Non-governmental Development Policies in Namibia and Their Impact on the Reproduction of Ju|'hoan Social Equity'. Paper presented at the Annual Meeting of the American Anthropological Association, Washington, DC, 15 November 1995.

Biesele, Megan, and Robert K. Hitchcock. 2011. *The Ju|'hoan San of Nyae Nyae and Namibian Independence: Development, Democracy and Indigenous Voices in Southern Africa.* New York: Berghan Books.

Coleman, James S. 1988. 'Social Capital in the Creation of Human Capital: Supplement: Organizations and Institutions: Sociological and Economic Approaches to the Analysis of Social Structure' supplement, *The American Journal of Sociology*, 94: S95–S120.

Diekmann, Ute, Maarit Thiem, Erik Dirkx and Jennifer Hays. 2014. *Scraping the pot: San in Namibia two Decades After Independence. Produced by the Legal Assistance Center and Desert Research Foundation of Namibia.* Windhoek: John Meinert Publishing.

Froerer, Peggy. 2012. 'Learning, Livelihoods, and Social Mobility: Valuing Girls' Education in Central India'. *Anthropology & Education Quarterly*, 43 (4): 344–57.

Gilbert, Jérémie. 2014. *Routledge Research in Human Rights law: Nomadic Peoples and Human Rights.* Florence: Routledge.

Glick Schiller, Nina, and Noel B. Salazar. 2013. 'Regimes of Mobility Across the Globe'. *Journal of Ethnic and Migration Studies*, 39 (2): 183–200.

Gordon, Robert, and Stuart Sholto-Douglas. 2000. *The Bushman Myth: The Making of a Namibian Underclass, 2nd edition.* Boulder: Westview Press.

Government of the Republic of Namibia. Office of the Prime Minister San Development website, <http://www.sandevelopment.gov.na/index.htm> accessed 2 January 2015.

Government of the Republic of Namibia. 2000. *Traditional Authorities Act.* Act Number 25, 2000, 2456.

Guenther, Mathias. 1999. *Tricksters and Trancers: Bushman Religion and Society.* Bloomington: Indiana University Press.

Hays, Jennifer. 2002. '"We should learn as we go Ahead": Finding the way forward for the Nyae Nyae Village Schools Project'. *Perspectives in Education*, 20 (1): 123–39.

——2007. 'Education, Rights and Survival for the Nyae Nyae Ju|'hoansi: Illuminating Local and Global Discourses'. PhD Thesis. Albany: State University of New York (SUNY), 2007.

——2009. 'Steps Forward and new Challenges: Indigenous Communities and Mother-Tongue Education in Southern Africa'. *International Journal of Bilingual Education and Bilingualism*. 12 (4): 401–13.
——2016. *Owners of Learning: The Nyae Nyae Village Schools Over Twenty-Five Years.* Basel: Basler Afrika Bibligraphien.
Hays, Jennifer, Rodney Hopson and Willemien le Roux. 2010. 'Evaluation of the Namas-supported San Education Project in Tsumkwe, Otjozondjupa, Namibia'. Evaluation Report. Namibia Association of Norway/NAMAS.
Ingold, Tim, 2000. *The Perception of the Environment. Essays in Livelihood, Dwelling and Skill.* London and New York: Routledge.
Kiema, Kuela. 2010. *Tears for my Land. A Social History of the Kua of the Central Kalahari Game Reserve, Tcamnqoo.* Gaborone: Mmegi Publishing House.
Lee, Richard. 1969. 'Eating Christmas in the Kalahari'. In *Natural History*, 14–22 December: 60–4.
——2013. *The Dobe Ju|'hoansi*. Belmont: Wadsworth.
Mitchell, Peter J. 2010. 'Genetics and Southern African Prehistory: An Archaeological Views'. Journal of *Anthropological Sciences*, 88: 73–92.
——2013. 'Southern African Hunter-Gatherers of the Last 25 000 Years'. *The Oxford Handbook of African Archaeology*. P. Mitchel and P. Lane (eds). Oxford: Oxford University Press.
Ninkova, Velina, forthcoming. 'Now That we Have Development: Change, Resilience and Education for the Omaheke Ju|'hoansi'. PhD thesis. University of Tromsø, 2017.
——2010. 'Under the big Tree'. Master's thesis. University of Tromsø, 2010.
Portisch, Anna. 2012. '"Like Unbroken Cream": Education and Livelihoods Among the Kazakh of Western Mongolia'. *Anthropology & Education Quarterly*, 43 (4): 385–99.
Salzman, Philip Carl. 2002. 'Pastoral Nomads: Some General Observations Based on Research in Iran'. *Journal of Anthropological Research*, 58 (2): 245–64.
Smith, Andrew B. 2001. 'Ethnohistory and Archaeology of the Ju|'hoansi Bushmen'. *African Study Monographs*, Supplement 26: 15–25.
Suzman, James. 2000. *'Things From the Bush': A Contemporary History of the Omaheke Bushmen. Basel Namibia Studies Series 5.* Basel: Schlettwein Publishing.
——2001. *Regional Assessment of the Status of the San in Southern Africa. Series of 5 reports.* Windhoek: Legal Assistance Centre.
Sylvain, Renee. 2001. 'Bushmen, Boers and Baaskap: Patriarchy and Paternalism on Afrikaner Farms in the Omaheke Region, Namibia'. *Journal of Southern African Studies*, 27 (4): 717–37.

Toren, Christina. 1993. 'Making History: The Significance of Childhood Cognition for a Comparative Anthropology of Mind'. *Man*, 28 (3): 461–78.
Van Rooyen, Piet H., and Peter Reiner. 1995. *Gobabis. Brief History of the Town and Region.* Gobabis: Municipality of Gobabis.
Veber, Hanne, Jens Dahl, Fiona Wilson and Espen Wæhle. 1993. '… Never Drink From the Same Cup'. Proceedings of the Conference on Indigenous Peoples in Africa. Tune, Denmark.
Wiessner, Polly. 2002. 'Hunting, Healing and Hxaro Exchange: A Long-term Perspective on !Kung (Ju|'hoansi) Large-game Hunting'. *Evolution and Human Behavior*, 23 (6): 407–36.
Willerslev, Rane. 2009. *On the run in Siberia*. Minneapolis: University of Minnesota Press.

ARNE ALEKSEJ PERMINOW

5 Moving Moorings, Nurturing Flows: Scales of Tongan Mobilities

> The food laid out on this table is the 'farewell meal' (*kai fakamavae*, literally 'food/eating of separation') for the 'foreigner' (*muli*) who has come here from afar to become accustomed to the 'manner of the land' (*anga fakafonua*). He shall return to his land far away, but we have lived together here on Kotu and he has become familiar with the people of Kotu. I believe that he shall remember this meal in the future. We have (all) 'mingled' (*'oku tau feohi*) during the time of his stay. We have 'helped one another' (*fetokoniaki*). This is why I think that this man shall not go abroad to disappear altogether from this land. There are such things as the mail and the telephone. Although he goes to live very far away the door is still open.
>
> — HEAMASI KOLA, TONGA (1991)

The farewell speech delivered by the head of the household I had been staying in during fieldwork on Kotu Island in Ha'apai, Tonga in both 1986 and 1991 (Perminow 1993a, 1996) was a not too subtle reminder that our relationship of 'mutual love/compassion' (*fe'ofa'aki*), 'mutual assistance (*fetokoniaki*) and 'mutual responsiveness' (*feongoaki*) should not end abruptly. That the obligation to keep on giving and receiving 'gifts' (*me'a 'ofa*, literally 'things of love') between us ought to endure separation. I do not believe that his words transformed me from an 'outsider' to an 'insider' with regard to cultural perspectives or local sociality. On the contrary, it was probably my persisting outsider qualities both in terms of kin relatedness and assumed social values that prompted him to be unusually clear about how relatedness may be established and made to endure.

An enduring involvement with a few Tongan families over the last three decades in Tonga as well as in New Zealand is part of the ethnographic foundation for the analysis. I shall use this long-term involvement to explore the relationship between, on the one hand, small-scale movements or flows that constitute persons and relationships within the course of everyday events in Tonga and, on the other, experiences, dilemmas, and tensions related to life-shaping 'trajectories of movements' (Salazar and Smart 2011, vi) developing over time as people seek careers and opportunities, that is, at a large scale, border-crossing mobilities of Tongans 'looking for a (good) life' (*kumi mo'ui*) through overseas migration. With Salazar I explore the value of 'taking on' mobility in a way that utilizes 'the ethnographic sensibilities of fieldworkers who learn about mobility while studying other processes and issues [...] and insist on attention to other dynamics that might not be considered if the focus is first and last on (im)mobility as such' (Salazar 2011, v; see also Chapter 6 in this volume). The bulk of the ethnography presented here was produced more than twenty years ago, not to learn about mobility or migration, but to understand dynamics of local sociality embedded in the multiple mobilities of mutual engagements in everyday life. Like Salazar then, I would argue that precisely an understanding of characteristic dynamics of local sociality, that 'might not be considered if the focus is first and last on (im)mobility as such' (Salazar and Smart 2011, v) may illuminate enduring mechanisms that contribute to constitute persons and relatedness as people move through life and beyond the limits of village, region, or nation. Not, of course, without friction (Tsing 2005) and modification, but still as essential means of coping with new challenges in novel social circumstances. My exploration thus represents an attempt to 'take on mobility' without letting the concern with large-scale border-crossing mobility lead to '[loosing] sight of the continued importance of place-based practices and modes of consciousness for the (re)production of cultures and societies' (Friedman 2002, cf. Salazar and Smart 2011, iv). Too often perhaps the dynamics of large-scale border-crossing mobility tends to be understood and analysed in contrast to a quite stationary normality (Larsen in this volume). The multiple movements of people and things, which are so characteristic of Pacific histories and societies challenge such a vision of stationary normality. I would argue that the normality of mobility makes Polynesian sociality a suitable field for exploring and illuminating how movements of substances, persons, and things may be interrelated and socially constitutive. In other words, I explore the potential of a focus on mobility to bring the dynamics of border-crossing migration within the same analytical framework as the multiple movements or changes of place, which created persons and relationships in the course of everyday socializing in Tonga during my fieldworks in Ha'apai in the last decades of the twentieth century.

My host's emphasis, in his farewell speech quoted above, on how shared experiences and mutual acts of 'love/compassion' (*'ofa*) are both caused by and result in social proximity reflects a core idea about social interaction in general and a strong ideal for kin relations in particular among Tongans. Tongan discourses on relatedness are replete with constructions of sharing and mutuality, which are linguistically achieved by prefixing and suffixing just about any word with *fe-* and *-'aki*. Thus the word *tokoni*, signifying 'help/assistance' can be turned into *fe-tokoni-'aki* signifying 'mutually helping one another'. The word *'ofa*, signifying 'love/compassion' can be turned into *fe-'ofa-'aki* signifying 'mutually feeling and acting with love/compassion toward one another'. And similarly, to mention just a few mutualizing constructions, *'ilo* or 'knowledge' can be turned into *fe-'ilo-'aki* signifying 'to recognize one another *tauhi* or 'care for' can be turned into *fe-tauhi-'aki* signifying 'to care for one another' and *hingoa* or 'name' can be mutualized as *fe-hingoa-'aki* signifying 'to name after one another'.

I will argue that an acute attentiveness to the socially constitutive force of sharing between people, be it material in the forms of things/substances or immaterial in the form of words, knowledge, and attentiveness is fundamental to Tongan sociability. I shall also argue that although such a sociability, deeply embedded in routines of behaviour and communication, is hardly the first to be left behind when people move overseas. However, the ideas and ideals underpinning such a sociability must of necessity rub up against a different scenario of challenges and opportunities when people's life trajectories move their moorings across new seas. Thus I would argue that other tensions and dilemmas develop when both the social environment and the distance across which people relate change dramatically through border-crossing mobility. But before doing so it is necessary to offer a brief introduction to the longer history of Polynesian mobility involved in bringing people to, between, and away from their Pacific islands.

Histories of Polynesians on the move

Human settlement on the isles of Polynesia, what some historians of culture have referred to as 'Remote Oceania' as opposed to 'Near Oceania' further west (Kirch 1997), shows that human mobility in the form of large-scale migration over vast distances has played a fundamental role in the past of Polynesians. By 3,000 BP, Austronesian-speaking seafarers appear to have reached and settled in Tonga, Samoa, and Fiji. This triangle is considered to be the oceanic space in which Proto-Polynesian language and culture developed over the next 1,500 years during which most of the so-called Polynesian Outliers to the west such as Futuna and 'Uvea (Wallis Island) were also colonized. Both linguistic similarities, similarities in material culture and oral traditions confirm that there must have been persistent and frequent movements in terms of flows of things, people, and concepts between the islands of this original Polynesian triangle. Some 1,400 years ago, large-scale Polynesian mobility brought people deeper into Remote Oceania as seafarers from the Tonga-Samoa-Fiji triangle explored the seas to the east and north and migrated to the uninhabited islands of Marquesas, Hawai'i, and Tahiti and finally from there south to Aotearoa-New Zealand, which was colonized approximately 700 years ago. The high level of boat building and navigational skills, rediscovered in the last decades of the twentieth century by the Polynesian Voyaging Society (Finney 1976), as well as the relative cultural and linguistic homogeneity, which the European explorers encountered in the eighteenth century, clearly attest to the enduring historical significance of inter-island movement within the region. The opening, however, of Polynesia to a larger world through European contact and colonization apparently reduced rather than increased Polynesian people's own large-scale mobility. In his work exploring Tongan modernity, Nico Besnier (2011) describes what Schiller and Salazar (2013) called a change in the 'regime of mobility' (Schiller and Salazar 2013) that accompanied the development of Tonga as a nation-state under British protection and influence:

> Throughout prehistory, Tongans were great travellers [...] Tongans travelled for many different reasons, including establishing new settlements [...] acquiring objects

> such as Fijian weapons [...] Samoan fine mats [...] borrowing dances from ʻUvea and Futuna; getting tattooed by Samoan specialists; seeking spouses and other kinds of alliances [...] [But] The early modern period represents a major break in that it was a period of unusual sedentariness, during which people were grounded more than ever before, their movements constrained by colonial and early state restrictions on ocean travel. (Besnier 2011, 37–8)

And this is clearly the case when it comes to large-scale, long-distance, and border-crossing mobility. Thus this period was characterized by the rapid decline of skills and knowledge related to the building and navigation of ocean-going sailing canoes, as long-distance voyaging became regulated by law and monopolized by Europeans. However, there is little doubt that inter-island mobility on a lesser scale within and between the island groups of Tonga remained a fundamental mode of securing food, artefacts, knowledge, spouses, and alliances in early modernity in Tonga as elsewhere in Polynesia. I have elsewhere shown how on the smaller islands in the periphery of Tonga, a comprehensive bilateral incest taboo between same generation relatives tend to make people move outside their home island to find eligible spouses (Perminow 1993a). I have also demonstrated how urbanization in general and an increased emphasis on the value of formal education and what is perceived as a good life has transformed a regime of mobility that moved people between islands in the periphery towards one that moves people between periphery and centre (Perminow 1993b). Thus it is mainly by equating mobility with large-scale border-crossing mobility that Besnier's characterization of the early modern period as 'a period of unusual sedentariness' (Besnier 2011, 37–8) is reasonable. In terms of peoples' incentives for moving as a strategy of securing a livelihood and a good life (if not in terms of its consequences), I would agree with Schiller and Salazar (2013) supported by King and Skeldon (2010) that there is little reason to 'side line' or ignore internal migration by maintaining an absolute 'division between internal and international migration' (Schiller and Salazar 2011, 184). Besnier is right, however, that it makes a significant difference whether the mobility has a scale limited to 'an island world connected by kinship' (Besnier 2011, 38), or whether people move far beyond what ʻEpeli Hauʻofa called 'Our Sea of Islands' (Hauʻofa 1993). Thus, it clearly makes a difference whether people move between similar kinds of places or whether they move between fundamentally different kinds of places which over time set those who move on and those who stay behind on quite different life trajectories. So, although it seems fair to argue that mobility has continuously been a basic mode of coping for Polynesians, the fact that over the last fifty years or so a steeply increasing percentage of people have moved overseas to settle in New Zealand, Australia, and the US, makes a difference. According to Besnier (2009, 222), perhaps as much as two thirds of people identifying as Tongans have moved outside Tonga and today the largest number of Polynesians (more than 50,000 of which are Tongans) in any single city live in Auckland, New Zealand.

It is also clear that in Tonga income generated by border-crossing mobility constitutes a very substantial flow of cash, which supports the increasing proportion of islanders who have moved from a village setting on small islands in the periphery of Tonga to the cash dominated urban setting around the Tongan capital, Nuku'alofa. Remittances take a variety of forms in addition to monetary transfers such as: 'clothing, household goods and furniture, building materials, tools, vehicles' (Lee 2003, 31) and also container-loads of articles for second hand sale in local flea-markets, making accurate figures hard to give. However, Lee documents that remittances were an absolutely essential means of coping for many Tongans in the 1990s, thirty years after border-crossing migration took off:

> To give an indication of just how significant remittances are to Tonga, from 1989 to 1990 the officially recorded remittances to Tonga totaled A$43.9 million, or 59.6 percent of Tonga's gross domestic product – and this figure represents only a portion of actual remittances. (Lee 2003, 32)

This heavy reliance on remittance has created concern among researchers as well as among people remaining in Tonga about how and to what extend migrants are prepared to continue remitting to their kin at home. Lee's findings appear to support the 'remittance-decay hypothesis, which claims that remittances decline when a migrant has been overseas for some time' (35), as close relatives follow kin overseas out of Tonga. Her findings also indicate that many second-generation migrants born overseas appear hostile toward the idea of remitting to distant kin remaining in Tonga. Remittances, it appears, must continually be fuelled by new migrants to flow strongly.

My own findings imply that in the first decade of this century the economic significance of another kind of border-crossing mobility increased markedly. The importance of temporary labour migration appears to have increased as homeland-based Tongans embrace opportunities to go to Australia and New Zealand for seasonal occupation, mainly in the fruit-picking industry. Although often criticized as social dumping in the countries where the work is done, most Tongans perceive it as a very welcome opportunity to come by cash that is otherwise quite hard to acquire in Tonga. In fact, several of my informants in Tonga, even those with quite close kin overseas, express low expectations of migrants' abilities to both meet the cost of securing a livelihood overseas *and* cater for the needs of those back home. Thus a married couple who moved from a small island in Ha'apai to the capital Nuku'alofa at the turn of the twenty-first century, in 2012 emphasized how much more they benefited from one of their son's income as a seasonal fruit-picker in Australia, and their other son's income as a sailor, than from remittances from a third son who had recently migrated to New Zealand. They did not blame him, however, knowing full well that money does not grow on trees even overseas and that life there be quite costly. Reduced costs of travel have given many people first hand experiences with life overseas and familiarized them not only with opportunities but also challenges and costs of living overseas.

Low expectations with respect to migrated kinsfolks' ability to remit may partly also be caused by the global recession in the first decade of the twenty-first century, which affected the construction sector in which many Tongans were involved as temporary workers. In 2011 the manager of a shop in Tonga's capital Nuku'alofa, selling sausages and other meat imported from New Zealand, reminisced about the 'good old days' at the turn of the century when her business was booming and her shop was crowded by people buying expensive, imported goods for everyday consumption and compared it to the present slow business where purchases were basically reserved for feasts and special occasions.

I have argued that a strong emphasis on sharing and an attentiveness to what moves in the spaces between people is fundamental to Tongan sociability and an important part of the cultural baggage (in terms of ideas underpinning sociality) that moves with people overseas. In line with

Sahlins' perspective on how cultural concepts are constantly modified by 'the risk of the category in action' (Sahlins 1985, 11) one could argue that such ideas are put at 'risk [...] in novel historical circumstances' (144). I would note, however, that Sahlin's concept of 'risk' perhaps too strongly associates intercultural interaction with forms of acculturation in terms of the destruction of cultural categories in favour of their modification in continually changing circumstances. Thus it might be more apposite to approach the dynamics through which fundamental Tongan notions of relatedness rub up against novel circumstances in terms of Tsing's concept of friction where: 'Cultures are continually co-produced in the interactions I call 'friction': the awkward, unequal, unstable, and creative qualities of interconnection across difference' (Tsing 2005, 4). Thus Tongans moving moorings overseas and those remaining in a village setting in Tonga move in different 'zones of cultural friction' (xi), producing different dilemmas and challenges as well as different cultural configurations. But even if a strong emphasis on the value of sharing and an attentiveness to what moves in the spaces between people should endure in the novel circumstances overseas, the maintenance of social relations does not have to involve remitting to homeland kin. Thus Tongan sociability may perfectly well thrive and be contained within Polynesian worlds of urban New Zealand, Australia, and the US. Worlds that, to quote Besnier (2011, 39) are 'often invisible to the dominant group, where the churches are Tongan, businesses cater to Tongans and employ Tongans, and Tongans inhabit entire neighbourhoods'. And as people *kumi mo'ui* or 'looking for life' move across new borders a sociability of what I shall refer to as constitutive flows may indeed play itself out within a diaspora that bypasses Tonga altogether other than as an idea of origin and a locus for self-conscious ethnic identity.

Tauhi vaha'a: 'Nurturing the space (between)'

It is my argument that a focus on moving moorings, understood as changing attachment or replacing one relational bond with another and as acts of establishing and maintaining social bonds by 'nurturing the space' (*tauhi vahaʻa*) between people is useful for understanding both large-scale movements across long distances and the many flows, movements, and changes of place of everyday village life. I would argue that movements or flows of 'things' (substances, artefacts, persons, names) related to social re-attachments created a highly dynamic and fluid topography of social proximity in the course of everyday village interaction during my long-term fieldworks in Haʻapai in the last decades of the twentieth century. I shall argue that a general preoccupation with 'nurturing' or 'feeding' (*tauhi*) social spaces characterize many kinds of small-scale, everyday Polynesian mobilities and may be key to an understanding of quite fundamental and enduring dynamics of sociality among Tongans in Tonga and beyond. Thus I use a focus on flows and movements of people and substances to try to relate the many kinds of small-scale mobilities that may be discovered through an ethnography of the practices and experiences of daily interaction to large-scale kinds of movements and flows transcending the local as well as the national.

This ethnography of daily interaction was not produced with the specific aim of relating characteristics of small-scale, everyday village mobilities to the dynamics of large-scale, border-crossing mobilities over time. However, I hold that such ethnography is essential in identifying general dynamics of Tongan relatedness that are neither left behind when people leave the village or Tonga altogether nor rendered insignificant as life careers develop overseas. Thus I return to my earlier ethnography of everyday village sociality in Tonga, exploring the creation and maintenance of social bonds through acts of 'feeding' or 'nurturing' social spaces (see Perminow 1997). I shall also use ethnography of attitudes and practices related to 'Tongan food' (*kai fakatonga*) and 'Tongan wealth' (*koloa fakatonga*), produced through three recent field visits among young migrants to Auckland who were raised in the small island community in Tonga where I started doing fieldwork three decades ago. My village ethnography is based on much thicker descriptions than my data from the migrant setting, which is much more anecdotal. Thus the village ethnography is the most sensible point of departure for an exploration of the significance of everyday movements or flows of substances, persons, and things for sociability.

Several students of Tongan society and culture have in recent years used the Tongan expression *tauhi vā* or *tauhi vahaʻa* to approach the characteristics of establishing and maintaining social relationships in general among Tongans and the active and dynamic aspect of kinship relations in particular. The expression means, literally, 'nurture/care for the space between' and has been used by scholars to approach the significance of reciprocity and networking in the reproduction of Tongan kin relations (see Ka'ili 2005; Leslie 2007; Poltorak 2007; Thaman 2008). My own findings confirm the centrality of such nurture/care in Tongan sociality but indicate, however, that the expression may also be used to refer to a particular mode of appearing/behaving in public that does not intrude on the space that separates people; that is, acting with decorum and respect. Thus, for some of my Tongan informants, *tauhi vaha'a* was not necessarily what one does to feed a kin relationship but what one refrained from doing out of 'respect' (*faka'apa'apa*) or 'love/compassion' (*'ofa*). In this sense, one older informant living in Auckland, translated it as 'to keep the peace' in public interaction. Contrasting the communicative style of kava drinking[1] in kava clubs in Auckland and informal sessions of kava drinking in a Tongan village setting, Tongan linguist Mele Taumoefolau (personal communication, March 2011) claimed that in the latter situation *tauhi vaha'a*, or 'caring for the space which separates the drinkers is so strong it is almost tangible' (Perminow 2015).

'*Vā*' or '*Vaha'a*' is a term referring to the space/void between any two entities regardless of scale and whether the entities are people, objects, or landmasses. It can be used to refer to the void or deep space between the earth and the moon, the deep sea between two landmasses, the open ocean separating Tonga from New Zealand overseas, or the distance between groups of people or between one person and the next. Thus the *Vā* or *Vaha'a* refers to the space in-between that separates things but also potentially connects them. In its sense of paying attention to decorum the use of the term 'nurturing/caring for the space' illustrates how not only material substances but also words and visual appearances are perceived to issue

1 *Kava* is a shrub in the pepper plant family (*piper methysticum*) with a mild psychoactive effect used in Tonga and elsewhere in the Pacific in both recreational and ceremonial contexts. For many Tongan men overseas its consumption in kava-clubs associated with specific local communities in Tonga remains a part of their weekly routines.

forth and flow in social spaces. The other part of the expression, '*tauhi*' signifies 'to feed', 'to nurture', 'to nurse', 'to tend' and 'to care for'. *Tauhi* is tending to your garden by removing weeds and by watering it. It is what you do to assist and care for people who are getting too old and weak to cope on their own. It is what you do to nurse people back to health when they are ill. And most of all it is what you do to nurture and feed infants so that they become big, strong, and 'healthy' persons (*sino* lit. signifying 'body' and 'fat'). *Tauhi* is strongly associated with motherhood – the mother's brother and the mother's side in general in Tonga. People rely on the mother, the mother's side, and maternal uncle in particular to nurture and support them. The father, the father's side, and the father's sister on the other hand, are ideally strict and demanding and should be expected to chastise you and teach you to act with respect and decorum. *Tauhi* may be described as the primary component of child care in the sense of being the only acknowledged ingredient in the first phases of childrearing until weaning. After weaning, which often occurs when the infant is between seven and twelve months old, *tauhi* is complemented with the other main component of Tongan childrearing known as *ngaahi/ngaohi*. *Ngaohi* signifies 'to put together', 'to build', 'to compose', 'to construct', which involves instruction/commands combined with generous doses of discipline, aimed at making persons *poto* or competent and which is strongly associated with fatherhood, the father's sister, and the father's side in general in Tonga (see Morton 1996; Perminow 1993a; Perminow 1996 for analyses of ideals and practices of Tongan childrearing). I would argue that the primary significance of *tauhi* in ideas about childrearing and a very strong emphasis on the strength of the primary attachment between mother and infant makes breast-feeding an act of *tauhi par excellence* in Tonga. I would further argue that quite enduring notions and practices related to the naval cord as a primal bond and an enduring source of personal well-being throughout life may be used as a further indication of the perceived significance of this primary mother-child attachment. All Tongan parents I have asked have claimed that they take care of the placenta (*fonua*) and the naval cord, often burying it at the roots of a tree in their town allotment. People say that it should be put in a dry/warm place rather than in a wet/cold place in order for the person to remain in good health throughout life. I have thus argued elsewhere (Perminow 2011) that the very obvious and explicit cultural emphasis of the higher status of father, father's side, and 'father's sister' (*mehekitanga*) in particular, which has been identified by several analysts of gender relations in Tonga (see Besnier 2004, 8), is accompanied by a more implicit understanding of a 'natural' strength of the bond with the mother, the mother's side and the ever-supporting 'mother's brother' (*fa'ētangata*, literally 'mother-man') (Perminow 2011). Discussions about the causes of qualitative differences in the raw materials for wood-carving with the Tongan carver Fe'ao Fehoko, with whom I worked as an apprentice in 2004, may illuminate the perceived natural strength and significance of this bond:

> This part [of the tree] is harder because it is 'in front/before' (*mu'a*), it leads the way and the other parts 'follow in its wake' (*mui mai*). It is just like the stomach which is the most important part of the human body. It is the preceding part of the body because in order to be strong and to think well you first have to eat. Just like the stomach the preceding part of the tree is the part which is closest to its 'place of origin'.

He used the term *fonua*, which signifies 'place of origin', 'land', 'territory', 'basis', as well as 'placenta' (Churchward 1959). Thus he elaborated how the stomach's importance as the seat of vitality and strength was caused by its connection through the naval cord and the placenta within the body of the mother as both origin point and medium of growth. He went on to emphasize how the mother and the mother's side of the family as the original source of the human body and its well-being was the one to rely on for 'assistance' (*tokoni*) and 'compassionate love' (*'ofa*). He referenced the aforementioned enduring practice of cleaning the afterbirth and burying it in a dry place in order to ensure good health throughout life, to emphasize the significance of this bond to the mother as the person's place of origin.

To 'nurture the space between' may profitably be understood, I would argue, as making an effort to keep or turn spaces that separate into spaces that connect persons and groups as strongly as flows of blood through the naval cord and flows of milk from mother to child. This sociability of nurturing spaces may be understood, then, as one in which primary, 'natural' bonds are transformed to and lend strength to other social bonds that are thus 'secondary' but also more highly valued in a system of cultural evaluation that generally prizes cultivation above things in states of nature. To 'nurture the space between' may be understood to involve paying attention to, establishing, and maintaining movements or flows that nurture and order the spaces between persons and groups. Thus the Tongan description for living harmoniously within kin-groups, neighbourhoods, or churches is *Vā lelei* (Leslie 2007) meaning literally 'good space' and implying that 'the nurturing' (*tauhi)* that turns spaces between social entities into something that unites rather than something that divides, is thriving.

Another metaphor – that of an 'open doorway' – to describe the significance of nurturing such spaces made by my host during my second fieldwork in Tonga 1992 illustrates precisely the perceived importance of making an effort to maintain a 'good space' through nurturing it by issuing things into that space even from afar.

Pusiaki: Everyday constitutive flows of persons and things

Above I claimed that the mother-child relationship epitomizes the constitutive potency of flows that connect and argued that this primary 'natural' kind of relatedness may be understood to feed and turn into secondary 'cultivated' kinds of relationships. If that is right then a focus on how 'cultivation' of children may be related to reattachment from mothers to others might be illuminating. In the last decades of the twentieth century, almost 40 per cent of the children and youth of Kotu Island in Tonga were staying with people other than their biological parents in more or less permanent and formalized arrangements of fostering or adoption. A focus on the characteristics and dynamics of such arrangements may be helpful in approaching precisely how moving and nurturing flows may be entangled with other cultural priorities in Tongan sociality.

The etymology of the term is unclear but '*pusiaki*' appears to refer to any person reared by others than biological parents whether in the form of adoption as a formalized, permanent shift of social attachment or in the form of a variety of informal fostering arrangements of varying permanence.

Some sources report that Tongans in earlier times may have distinguished between social reattachments known as *tauhi* ('to nurture, care for') relations of fostering, and *ohi* ('a transplanted plantain') relations of permanent adoption (Beaglehole 1944, 71; Gifford 1929, 26–7; Urbanowicz 1973). No one I have asked on Kotu knew this distinction though, and there is a good chance that it may have been produced to satisfy a European demand for categories of social attachment resembling our own categorical distinction between fosterage and adoption. Elsewhere I have argued that a quite literal sense of the Latin term *ad-optione* pinpoints the essential characteristics of *pusiaki* relations in identifying optionality as its defining quality (Perminow 1996, 330). Thus my ethnography shows that shifts of attachment producing a *pusiaki* relation between child-givers, child-takers, and the child itself was something that must be opted for by both giver and taker and sometimes the child itself. Also it was mostly a relation that would break down if either child-giver, child-taker or the child itself opted out of it and which quite quickly dropped out of local common memory.

Fostering arrangements comes about in a variety of ways and in different stages of a child's life. On Kotu in the 1990s a considerable percentage of *pusiakis* were adopted when they were between seven and twelve months. A few were adopted when they were between one and ten years old, while a considerable percentage was adopted when they were eleven to twelve years old. It was mostly individuals rather than couples who adopted them and very often close kin among whom the father's sister of the adopted child stood out as the most common child-taker. According to some informants, adoption should ideally not take place until the child showed signs of having developed a 'human mind' (*atamai tangata*) – they claimed that very young infants who do not respond to sudden movements in front of the face were 'still of indifferent mind' (*kei atamai noa*). Later, when they respond to sham thrusts in front of the face, they are 'still of animal mind' (*kei atamai manu*). And finally, when they start to cry if the mother leaves their sight, they are 'already of human mind' (*'osi atamai tangata*). Showing the first awareness of the value of a particular social bond then, the child would be considered ready for *ngaohi* or 'building' in addition to *tauhi* or 'nurturing' by complementing or severing that very first bond. This would, they claimed, be the time for physically turning the child out to face the world when sitting on the lap of the mother instead of facing inward, the time for weaning, preferably by leaving the child with the father's sister for a time, and a good time for fostering it out, ideally also to the father's sister who in terms of Tongan kin ideology is a strict relative that you should consistently treat respectfully and defer to (*faka'apa'apa*).

Most fostering arrangements did not involve large-scale movement but instead relocation of children to other homes in the same village. Parents nevertheless emphasized the pain and suffering of the mother at the severance of her close bond to her child. Thus they would typically say: 'When her child was sent away to be adopted she cried a lot for many days'. The most common age of adoption is just as the child approaches one year, has reached the developmental stage referred to as *atamai a tangata* or 'of human mind' and thus considered ready to be *ngaohi*, that is, 'assembled' or cultivated by learning how to behave in general and how to behave differently to different kinds of relatives in particular.

The second most common age of adoption appeared to coincide with another significant stage of becoming a cultivated person, namely at the transition between primary schooling on the island and secondary schooling outside the island at the age of twelve. Thus some of those referred to as *pusiakis* were adopted through an arrangement whereby a relative paid the school fees and supported with fish and agricultural produce other relatives with whom the child was staying during terms. Between terms such *pusiakis* would stay with these foster parents, helping them rather than their biological parents at least for the duration of their secondary schooling.

Finally the reattachment of children also appeared to be entangled with notions of building, cultivation, and the forming of the person through practices of naming. This was hard to miss when I conducted a village census on who had named children and after whom they had been named in 1991: as I interviewed the household head of the homestead next to the one in which we were staying, I knew perfectly well that their youngest son born a few months after my first stay on Kotu four years earlier had been named after my own son who had accompanied me to the field. Ever since our return to Kotu in 1991 our neighbours had encouraged their four-year-old son to spend time with his *palangi* ('foreigner') namesake and to sit on his lap in our homestead. They often pointed out how they had 'the same face' (*mata tatau*), behaved in the same manner, and had the 'same disposition' (*anga tatau*), as well as similar food preferences. Thus they claimed that he was just as *pau'u* or 'naughty' as his namesake had been during the first visit when he was of the same age. They also pointed out that he had a particular liking for imported *pālangi* or 'foreign' food. So, evidently the fact of his naming was presented as something contributing to shaping him. Thus although I knew who he was named after and realized that this was perceived to form his character, I didn't know who was responsible for his naming. So I asked his father about it. He seemed quite surprised by my question but quickly answered: 'Well, it was you!' This was news to me and I asked him to elaborate. 'Shortly after my son was born in 1987, when you had gone back abroad', he said, 'you sent a letter to the "senior head" (*ulumotu'a*) of our (kin) group and the "guardian of the land" (*motu'a tauhi fonua*). You told him that the boy should be named after your son. He agreed because you and your son spent much time with us when you were here and your son often ate and slept with us. He didn't want the name to be lost from Kotu. So he told me about your wish and I thought it was a good thing too'. Unable to recollect this correspondence, I merely nodded my head, but realizing my perceived involvement in exchanges contributing to the 'building' of this child, I asked whether it would be appropriate for me to help with costs related to the boy's schooling. 'That would', the father responded 'be quite appropriate and a very good thing'.

Following the clues from this encounter, I discovered that naming was related to childrearing both by children 'inheriting' qualities from the ones after whom they were named and by opening up a possibility for a future attachment of specific children to specific name-givers perceived as particularly resourceful and thus as someone who might contribute to the child's success in 'looking for a (good) life' of their own. For example, people who had succeeded in achieving prominent positions in church organizations would quite often be asked to name a child and would by doing so take on an enduring responsibility or sponsorship. Such obligations might be intermittent but might also develop into a temporary or permanent fostering arrangement. The perceived significance and endurance of names as one of the 'things' issuing forth and into the spaces between people may be illuminated by fast-forwarding thirteen years to another encounter in 2004: trying to learn how to carve a proper Tongan headrest as an apprentice at the workshop of a Tongan woodcarver just outside Nuku'alofa, I felt the need to examine old Tongan headrests. This was easier said than done since most such headrests have ended up in museum collections outside of Tonga. Recollecting, however, that old Amini, living a few homes down from us on Kotu in 1986 and 1991, used to take his midday naps resting his head on a well-worn Tongan headrest, I decided to try and find out what had happened to it in the meantime so that I could examine it in detail. I went back to Kotu only to find out that Amini's sister had brought it with her, away from Kotu, when Amini passed away in 1992. When she in turn passed away a few years later it finally made its way to Amini's younger brother Hateni, who worked as a church steward in Nuku'alofa. He was in his late seventies and as I entered his house he was having a nap after his day's work in the bush, resting his head on it. I told him about my apprenticeship and my need to examine the headrest. 'How is your son, the one the son of my mother's brother's son was named after on Kotu?' he asked. 'He is fine', I said. 'He has just completed his military service in the King's Guard in Norway'. 'I remember' he went on,

> how content (*lata*) he was on Kotu and how familiar he was with the people there, how fond he was of sugarcane and how close you both were with the home of my mother's brother's son. And his namesake ... so naughty ... just like him and preferring *pālangi* food too! And now he is in Form 4 at 'Atele College!

We were sitting on the floor facing one another. He straightened up, becoming more formal in bearing and lifted up the headrest with both his hands and held it out saying,

> I can see that this is a very important thing for you because of your work. It is really old. I cannot remember a time when it was not there and it was already old when I saw my father resting his head on it when I was still small on Kotu. My father's father rested his head on it before that and my eldest brother before me. It has made it for more than a hundred years! It will be a good thing if you take it and bring it to your land so that your son can have it after you to remember his Kotu family from the other side of the world. That will be a beautiful thing!

I was quite moved by his unexpected little speech but also quite uncomfortable at the prospect of moving yet another old artefact out of Tonga. I asked if it would not be better to leave it for his own children. 'This is much more important for you' he said,

> than for them. They are married and their wives do not 'care for' (*tauhi*) other than the 'heirlooms' (*koloa tuku fakaholo*) from their own sides so they will neglect it and it will be broken or lost. Also my daughter is unmarried and already old so it will not pass on that way.

When I still hesitated to accept the headrest, he lifted it up a bit higher and with tears in his eyes and a voice thick with emotion he simply said, '*Koe 'ofa*!', meaning literally, 'This is love/compassion!'

And also I would argue, it was an act of using a 'thing' as strongly united as a child with her mother or a name with a person to feed into and nurture social spaces established by prior acts of sharing and foreseeing a future of feeding those spaces. Thus it was not too surprising when my son's namesake four years later named his first-born daughter after my eldest daughter and asked for financial help when he won the lottery by which the annual quota of Tongan migrants to New Zealand was allotted in 2010. I was somewhat more surprised, however, when I visited him in Auckland in 2012 and he asked me to pay for the airfare for and escort the four-year-old namesake of my daughter back to her grandparents in Tonga as his marriage was breaking up. On the basis of my current argument, however, that names are potentially forceful things entangled with other things issuing forth and flowing in 'the space between' to create and recreate social proximity, perhaps I shouldn't have been.

In village interaction the numerous *pusiakis* appeared to play a key role in intensifying the 'mutual exchange of help, food and money' (*fetokoniaki*) in the space between child-giver and child-taker. Thus *pusiakis* seemed both the ideal 'beggars' for things and services and bringers of messages and small gifts of food between the household of child-givers and child-takers. They were very often the concrete vessels traversing the 'space in between' to connect the two. As long as all the three parties were not too dissatisfied, as long as exchanges were flowing strongly and were not too unbalanced and as long as the child remained 'contented' or *lata* the space in between appeared to remain a *vā lelei*; a 'good space'. But when one of the parties felt a lack of 'love/compassion' (*'ofa*), a reluctance to help out or if the child felt maltreated or became 'homesick/discontented' or *ta'elata*, experience feelings of anger or *'ita*, then this would interrupt the flows that produces a 'good space in between'.

The following case developing over the decade from the early 1980s to the early 1990s illuminates the fluid quality of adoption arrangements and the social dynamics involved in the maintenance of 'a good space' between child-giver, child-taker, and the child given. Around 1983, when the boy was five years old, he started to 'go a lot' (*fa'a 'alu*) to the home of a cousin, three years older, who as an infant had been adopted into the home of a childless couple. He was said to have attached himself to this older friend. By 1986, when he was eight years old, he was recognized by all the villagers, including his biological parents, as a *pusiaki* in the friend's home. His adoptive father was quite well off, and people felt that being adopted out of a relatively poor household with many mouths to feed into one with just two *pusiakis* would improve the boy's chances of getting a secondary education. Thus his biological parents were pleased with his move and contributed to his upkeep by sending with him fish and cooked food when he stopped by and by responding to and 'soliciting help' (*kole*) in the course of everyday needs. At the end of 1986, however, the adopted mother of the two boys adopted a third child, a seven-month-old girl from her father's sister's daughter. From that day the boy and his elder foster brother were expected to stay around the home more than before to be at their foster mother's beck and call as she cared for the infant girl. In 1987, his elder foster brother moved out of the household to go to secondary school in Nuku'alofa, leaving all the domestic chores to the nine year old boy. Some said the boy grew 'discontent/homesick' after that and others said his biological father, seeing how his child was burdened with chores, began to doubt whether his foster parents would actually let him go to secondary school. 'This is why', one said, 'his biological father finally "welcomed" (*tali lelei*) him when the boy kept running "home" away from his foster parents'. The foster-father emphasized the pain and suffering of his wife when the foster son kept running home and criticized the biological father for 'receiving the son too readily' instead of sending him back to them. But in 1991 when he looked back at the events resulting in the boy's return to his native home, he concluded that if they had tried to force the boy to stay against his will it would have 'destroyed the boy's mind'. He also said that the 'strong fellowship' (*feohi*) and 'mutual helping' (*fetokoniaki*) that had connected their home with that of the boy's biological parents' stopped when he finally returned to his native home during 1988. In 2004 none on Kotu remembered a time when the boy was a *pusiaki* but did recall that 'he used to go there a lot'.

Locally then *tauhi vaha'a* ('nurturing the space between') and the small-scale re-mooring of children are intimately interrelated. I have argued that child reattachments through adoption lend their socially constitutive force from a 'naturally' strong bond established through primary constitutive flows from mother to child, prenatally through naval chord and then through her breast. Some who have analysed the prevalence of fostering and adoption arrangement in Polynesia conclude that it demonstrates a weak bond between natural parents and their offspring in general and mother and child in particular (Carroll 1970). Contrary to this, my ethnography suggests that the common perception of the natural strength of this bond makes it socially constitutive to reattach children in arrangements of fostering and adoption.

Re-moving moorings, re-scaling flows

From my briefer but more recent encounters with Tongans who are moving overseas it is clear that Tongans overseas find the demands made by home-based kinsfolk quite a burden and some may even feel that 'Tonga only wants our money!' (Lee 2006). Kinsfolk at home may also find it increasingly hard to convince those who have moved oversea that they really need the 'Tongan wealth' and the 'Tongan food' that homeland kin have to offer. Thus as businesses catering to Pacific Islanders' needs flourish (Besnier 2009, 38; Small 1997, 70) and as the form of traditional ceremonial transforms to suit new environments (Perminow 2015) migrant communities appear to become increasingly self-sufficient. Particularly second generation migrants may be unwilling to 'carry the burdens' *(fua kavenga)* of nurturing the spaces between their closest family and more distant kin in Tonga. Some were quite hostile to the idea of using resources on traditional 'Tongan wealth' (*koloa*) of plaited mats and bark-cloth at ceremonial occasions. Thus in 2010, a New Zealand-born Tongan in his early twenties whose mother migrated from Kotu to New Zealand in the early 1980s said that he had grown up on 'Tongan rubbish', referring to the heaps of bark-cloth and plaited pandanus mats his mother and grandmother were hoarding under the matrasses of the beds in their home. He seemed to regard 'Tongan wealth' as useless, something that their homeland-based kin tried to push on them in order to get hold of their money. Even those less hostile and more 'into Tongan wealth' and sensitive to its usefulness at ceremonial occasions, sometimes accounted for their hordes of such valuables in terms of 'responding to solicitations' (*tali 'a e kole*) from relatives in Tonga and of 'helping them out' (*tokoni*). Thus one of the largest rooms of the home a Kotu woman in her early seventies who migrated to Auckland in the 1960s was full of 'Tongan wealth'. She described it as the result of her not being able to refuse when she was contacted by distant kin who were in need of quick cash in order to be able to fulfil obligations to contribute at quarterly or annual collects of money in their village churches in Tonga. Her daughter-in-law, who lived under the same roof, on the other hand, declared herself as 'not into Tongan wealth' and felt that the older generation's preoccupation with hoarding and presenting it in huge quantities at first birthdays, weddings, and funerals was useless and wasteful and would soon become a thing of the past. Clearly the willingness to divert resources from the spaces of the most immediate social surroundings in favour of spaces extending overseas back to Tonga varies considerably between first- and second-generation migrants. Apparently also attitudes to the constitutive value and force of constantly feeding or nurturing social spaces with 'Tongan valuables' may be undergoing significant change as it rubs up against a different configuration of challenges and opportunities overseas.

But what about what I, based on ethnography of the small-scale mobilities of everyday sociality in a village setting, have argued to be a core quality of Tongan sociability; namely the perceived constitutive potency of relinquishing what in Weiner's vocabulary would be inalienable possessions (Weiner 1992) intimately attached to persons and groups and thus allowing existing (primary) bonds to turn into new social bonds? Again, I propose, a focus is needed on how children move in large-scale, border-crossing spaces between people and how they are entangled with other things issuing into such spaces. Also I propose that a focus on tensions and changing attitudes related to such movements may illuminate how a Tongan sociability of nurturing spaces plays itself out in the new 'zones of cultural friction' (Tsing 2005, xi) in which Tongan migrants are 'seeking a life' (*kumi mo'ui*).

Children play a significant role in 'nurturing the space' (*tauhi vaha'a*) between those who 'seek a life' overseas and kinsfolk trying to cope in Tonga, albeit mostly quite close kin. Thus my own findings concur with Lee's findings in her examination of the Tongan diaspora communities of Melbourne (2003), which showed that pre-school children are often sent back home to be cared for by close kin until the children enter primary school. Also quite a few are sent back home for shorter or longer periods in order to be 'straightened out' by means that are unlawful in New Zealand. Apparently then, just as with regard to reattachments of children entangled with multiple small-scale mobilities in a village setting, such reattachments seem to happen quite early and quite late in childhood and to be related to enduring notions of becoming a well cultivated Tongan person. Lee documents how many Tongan parents in the Australian diaspora felt that it was important that their children spend their early childhood with relatives in Tonga (mostly quite close kin like grandparents or an aunt or uncle) to learn both the Tongan language and the 'Tongan manner' (*anga fakatonga*). She also documents the tendency of sending wayward teenagers who were becoming streetwise rather than 'educated' (*poto*) 'to be corrected' (*fakatonutonu*).

In 2010 one of my own informants in Auckland, born on Kotu in the late 1970s, described this strategy of childrearing as one of 'making straight' *fakatonutonu* something that is turning 'crooked' or *pikopiko*. He remembered how his own father had taken him and his cousin for a whole month of what he called 'hard labour' on the volcanic, inaccessible and 'difficult' (*faingata'a*) island of Tōfua, where most Kotu farmers have additional land. This was after the father had found out that they were not attending the education he was breaking his back to provide for them but they were getting drunk on homebrew instead. He described it as a sort of military camp where his father forced them up before the crack of dawn and worked them hard at the plantation until nightfall, serving them meals consisting mainly of *manioke* tubers and asking them: 'Is this the kind of life you want? For this is surly the life which awaits you if you drink homebrew instead of going to school!'

Lee (2003) emphasizes the significance of sending children back home for the development of their linguistic skills in Tongan, their familiarity with 'the Tongan manner' and their sense of Tongan identity. And this is clearly one of the considerations many Tongans in diaspora would site as an appropriate reason for sending their children back home. However, a relative lack of emphasis on making children speak Tongan, eat 'Tongan food' or act in the 'Tongan manner' characterized the Tongan homes in Auckland that I visited from 2010–12. Although many of their children had spent many of the years prior to entering school with relatives in Tonga and thus quite often spoke very little English when entering school, it was more or less expected of them that they would quickly turn away from speaking Tongan in favour of English. Although spoken to in Tongan by their parents, most second-generation school kids respond in English. Also it was not expected of school kids that they should have the same sort of food preferences as their parents. Thus parents would normally serve cereals, toast, jam, and eggs for the kids' breakfast while maintaining a diet much richer in the root crops on which they had grown up themselves. Finally, most Tongan parents in New Zealand seemed to take for granted that their children were confused about how to act appropriately towards different kinds of relatives. This lack of emphasis on acting and speaking Tongan in the course of everyday life, as well as that of sending wayward youth approaching the end of childhood overseas, not only to make them familiar with their cultural roots but sometimes to motivate them to work harder to qualify themselves for a life overseas, implies a deep ambivalence among parents about the usefulness of their cultural baggage for social and economic success overseas.

According to some informants 'seeking their life' in New Zealand, having custody of children whose natural parents are moving their moorings overseas increases substantially the flow of remittances in cash and kind for the duration of the fostering arrangement. It might appear then that when the space between a child-giver and a child-taker is small, like that between two neighbouring homesteads in a single village, intermediate, like that between the village where the child was born and raised and the town or city where she stays to attend secondary school, or really large, like the space separating overseas child-givers from child-takers in Tonga, movements of children both build upon, cause, and are accompanied by exchanges of things. My findings indicate that the sending of small children back to Tonga to be cared for by close kin there may have been not only about making them Tongan but about nurturing and maintaining the large space between overseas child-giver and homeland child-taker by acts of *fe'ofa'aki* or 'mutual love' and *fetokoniaki* or 'mutual help', which remain vital enactments of the core social value of *'ofa* or 'love/compassion' for most Tongans wherever they 'seek their life'.

Such an arrangement is potentially of mutual benefit to overseas parents and homeland kin by making it possible for those overseas to work while their children are small and by facilitating a flow of remittances back to caretakers in Tonga. Tensions and conflicts among overseas spouses related to choice of child caretaker in Tonga and what else, how often, and how much of it should be channelled into the space between them, however, indicate that some migrants understand such caretaker arrangements to be primarily in the interest of a caretaker who, to borrow Lee's formulation, may quite readily be perceived to 'only want our money' (Lee 2006). In particular such an approach to homeland-based caretakers' desire to share the burdens of childrearing appeared to dominate the understanding of in-laws' motives for fostering migrants' children, creating tensions and conflicts in marriages. Although united in feeling obliged to 'nurture the space' between Auckland and Tonga, spouses often appeared to be divided with regard to whose side should be nurtured. According to Lee's findings, women are generally more reliable remitters than men (Lee 2011), which could indicate that the wife's side is most often prioritized. The understanding of one of my informants who had migrated to Auckland in the late 1990s clearly reflects, however, a considerable dissatisfaction with having agreed to send their youngest daughter to be cared for by his wife's family in Tonga. He was fed up, he said, with the greed of his wife's mother back home and the constant solicitations for support based on her caring for their daughter. He was looking forward to the day his daughter would return to Auckland to start school and when he would be able to keep his wages out of the in-laws' grasp. Since his wife did not have a job outside the home but looked after their three other school kids, he felt that they had no benefit from the fostering arrangement at all.

In 2011, the extent of his resentment was made even clearer when there was a large first-birthday celebration on his wife's side of the family in New Zealand. His wife had spent all night with her sisters baking a dozen colourful cakes for the occasion. He took no part in the preparations and did not attend the feast, preferring to stay at home with his cousin who had just arrived from Tonga, both of them referring to in-laws as 'our enemy'. At my next visit in 2012, the marriage had broken up, his wife had moved out, the three school kids remaining with him, the fourth presumably still with his wife's family back in Tonga.

Marriage statistics among Pacific Islander migrants to New Zealand do not suggest that Tongan marriages break down more often than other marriages in New Zealand. On the contrary, statistics imply that divorces are uncommon. However, there may be a considerable disparity, for whatever reasons, between marital status and living arrangement. Thus spouses may 'live apart' (*mavae),* 'move in with new partners' (*nofo fakasuva*) and have children moving in the different spaces of their different sides in New Zealand or back in Tonga, without filing for a divorce. I suspect that what the marriage statistics more than anything else show is that it is unusual to file for divorce among Tongan migrants. So in 2012, when a Tongan woman in a conversation we had about divorce rates, claimed that 'divorce has become extremely common among Tongans', she might have been referring to *de facto* 'separations' (*mavae*) that do not readily show up in statistics.

Tongan marriages, as other marriages, break down for many reasons. But moving married life from very closely knit and transparent local communities affording husbands and wives the possibility to socialize on a range of arenas to much more loosely knit and less transparent communities in which homes are relatively insular and closed may make a significant difference. Thus the relative insularity of migrant homes, combined with a very strong identification of work at home as the wife's domain and work outside the home as the husband's domain, made the wives in the two marriages that were breaking up during my visit to Auckland in 2012 describe their homes like prisons and themselves as 'slaves' (*popula*). I suggest that this feeling of insularity might both increase the commitment to opening the doorway of the home to nurture the spaces between it and other homes and also contribute to tension and ever-present potential for conflicts about which spaces to nurture and what kind of valuables to keep within the boundaries of the home in the first place.

Conclusion

Reattachments of children for all or parts of their childhoods and the many choices and dispositions their parents must make with regard to what spaces to nurture, has great potentials to create strains and disagreement. Tongans who are moving their moorings overseas no doubt share with those who move closer to their home islands fundamental cultural notions about the socially constitutive significance of nurturing spaces by letting go of things that are strongly attached to them. I do think, however, that differences in scale of the *vaha'a*, or the 'space between', as well as differences in the social contexts of those letting go of such inalienable possessions make a difference. Thus the cultural assumptions or categories on which nurturing spaces are based rub up against different challenges and affordances when the space is nurtured overseas or across it increasing tensions, dilemmas, and conflicts related to whose side and which spaces to feed and how much and what kinds of possessions to feed it with. Also, as Sahlins pointed out long ago, arguing that cultural reproduction or change may profitably be understood as to sides of the same coin, cultural categories or fundamental notions about sociability, is always 'risked in action' and are perpetually modified as they are confronted and rub up against novel circumstances (Sahlins 1985, 144). More ethnography among Tongans moving overseas is needed, but preliminarily, I would suggest that a Tongan sociability based on the social significance of detachment, reattachment, mutual flows, and nurturing spaces may in the long run be finding itself between a rock and a hard place as people are moving overseas. Still, my argument has also been that the socially constitutive force of cultural notions about how relations are established and maintained through 'nurturing, feeding or caring for the space between' (*tauhi vaha'a*) is considerable wherever Tongans may be based. Such cultural notions and their capacity to establish and maintain social relations may most easily be discovered perhaps in the movements, flows, and changes of place of substances, people, and things in the course of everyday village life. However, the notions and their socially constitutive force are not left behind in the village nor at the crossing of the Tongan border – in fact, they know no boundaries.

Bibliography

Beaglehole, Earnest and Pearl. 1944. *Pangai, a Village in Tonga.* Wellington: Polynesian Society Memoir 18.

Besnier, Nico. 2004. 'The social production of abjection. Desire and silencing among transgender Tongans'. *Social Anthropology*, 12 (3): 395–401.

——2009. 'Modernity, Cosmopolitanism, and the Emergence of Middle Classes in Tonga'. *The Contemporary Pacific*, 21 (2): 215–62.

——2011. *On the Edge of the Global. Modern Anxieties in a Pacific Island Nation.* Stanford: Stanford University Press.

Carroll, Vern. 1970. *Adoption in Eastern Oceania.* Honolulu: University of Hawaii Press.

Churchward, C. Maxwell. 1959. *Tongan Dictionary.* Nuku'alofa: Government of Tonga Printing Press.

Finney, Ben. 1976. *Pacific Navigation and Voyaging.* Auckland: The Polynesian Society.

Friedman, Jonathan. 2002. 'From Roots to Routes: Tropes for Trippers'. *Anthropological Theory*, 2 (1): 21–36.

Gifford, Edward W. 1929. *Tongan Society*, Bernice P. Bishop Museum Bulletin 61, Honolulu.

Hau'ofa, 'Epeli 1994. 'Our Sea Of Islands'. *The Contemporary Pacific*, 6 (1): 148–61.

Ka'ili, Tevita. 2005. 'Tauhi Vaa: Nurturing Tongan Socio-spatial Ties in Maui and Beyond'. *The Contemporary Pacific*, 17 (1): 86–105.

King, Russel, and Skeldon, Ronald. 2010. 'Mind the gap! Integrating Approaches to Internal and International Migration'. *Journal of Ethnic and Migration Studies*, 36 (10): 1619–46.

Kirch, Patrick V. 1997. *The Lapita Peoples: Ancestors of the Oceanic World*. Cambridge: Blackwell.
Lee, Helen. 2003. *Tongans Oversea. Between two Shores*. Honolulu: University of Hawai'i Press.
——2006. 'Tonga Only Wants our Money: The Children of Tongan Migrants'. in *Globalisation, Governance and the Pacific Islands*, edited by Stewart Firth, 121–35. Canberra: ANU E Press.
——2011. 'Rethinking transnationalism through the second generation'. *The Australian Journal Of Anthropology*, 22 (3): 295–313.
Leslie, Heather E. Young. 2007. '... Like A Mat Being Woven'. *Pacific Arts* 1 (3/5): 115–27.
Morton, Lee Helen. 1996. *Becoming Tongan, an Ethnography of Childhood*. Honolulu: University of Hawai'i Press.
Perminow, Arne A. 1993a. *The Long Way Home. Dilemmas of Everyday Life in a Tongan Village*. Oslo: Scandinavian University Press.
——1993b. 'Between the Forest and the Big Lagoon: The Microecenomy of Kotu Island in the Kingdom of Tonga'. *Pacific Viewpoint*, 34 (2): 179–93.
——1997. *Moving Things of Love. An Ethnography of Constitutive Motions on Kotu Island in Tonga*. PhD thesis, University of Oslo.
——2011. 'It is a tree that fights: Engaging notions of qualitative difference in Tonga'. In *Identity Matters: Movement and Place*. I. Hoëm and R. Solsvik (eds), Occasional Papers, Volume 12. Oslo: Kon Tiki Museum, 111–23.
——2015. 'Food Presentations Moving Overseas: Ritual Aesthetics and Everyday Sociality in Tonga and among Tongan Migrants'. In *Objects and Imaginations; Perspectives on Materialization and Meaning*, edited by Øivind Fuglerud and Leon Wainwright. New York: Berghahn Books, 111–33.
Poltorak, Mike. 2007. 'Nemesis, Speaking, and Tauhi Vaha'a: Interdisciplinarity and the Truth of "Mental Illness" in Vava'u Tonga'. *Contemporary Pacific*, 19 (1): 1–36.
Schiller, Nina G., and Noel B. Salazar. 2013. 'Regimes of Mobility Across the Globe'. *Journal of Ethnic and Migration Studies*, 39 (2): 183–200.
Sahlins, Marshall. 1985. *Islands of History*. Chicago: University of Chicago Press.
Salazar, Noel B., and Alan Smart. 2011. 'Anthropological Takes on (Im)Mobility'. *Identities: Global Studies in Culture and Power*, 18 (6): i–ix.
Small, Cathy. 1997. *Voyages: From Tongan Village to American Suburbs*. Ithaca: Cornell University Press.
Thaman, Konai Helu. 2008. 'Nurturing Relationships and Honouring Responsibilities: A Pacific Perspective'. *International Review of Education*, 54 (3): 459–73.
Tsing, Anna. (2005). *Friction. An Ethnography of Global Connection*. New Jersey: Princeton University Press.

Urbanowicz, Charles F. 1973. 'Tongan Adoption Before the Constitution of 1975'. *Ethnohistory*, 20 (1): 109–24.
Weiner, Annette B. 1992. *Inalienable Possessions. The Paradox of Keeping-While-Giving*. Berkley: University of California Press.

JAN KETIL SIMONSEN

6 Kinship and Mobility in Urban Zambia (and Beyond)

Much mobility research has focused on the shape and practices of social networks in space-time. A point of concern has been the ways in which people sustain social relations across large geographical distances as people's networks expand and change through the life course and under different political and economic regimes (Bissel 2013; Glick Schiller and Salazar 2013). Key concept-metaphors in this line of inquiry are 'weak' and 'strong' ties. Whereas strong ties signifies relations of commitment and emotional nearness, conventionally regarded as being based on physical and social proximity, weak ties signifies the 'stranger' in Simmel's (1950) sense; a person who is nearby, but socially distant. From the point of view of more conventional social theory, increased mobility is seen as leading to a reduction of strong ties in people's individual networks and a proliferation of weak ties that are short-lived because '[w]eak ties seem suitable for mobile life styles; they represent *routes while strong ties implies roots*' (Larsen, Axhausen and Urry 2006, 262, emphasis added). Albeit rebutting such views, mobility researchers tend to treat social proximity in individual social networks as *requirements* and regard mobility primarily as practices, infrastructure, technology, and capital employed to maintain and reproduce *existing* relations (Bissel 2013). They neglect how mobility establishes new social relations and produce new forms of proximity (see also Chapter 1 in this volume). A complementary line of inquiry has thus been suggested, namely to examine the cultures, practices, and experiences of *forming* social relations among people on the move (Bissel 2013; Glick Schiller and Salazar 2013; Glick Schiller, Darieva and Gruner-Domic 2011).

In this chapter, I examine cultural practices among migrants and mobile persons in Zambia that sustain existing relations, as well as form new ones. More particularly, I examine mobility and frequencies of contact and exchange among Mambwe-speaking persons who after Independence in 1964 dispersed and settled all over Zambia for education and employment. The social network is seen from the perspective of a group of classificatory brothers who in a process chain migration settled in the capital Lusaka more than 1,000 km away from their home villages in Mbala District, Northern Province. Their households in Lusaka became mutually dependent for survival, but also a destination for rural Mambwe who for one reason or other visited or travelled through the capital.

After Independence from British colonial rule in 1964, the government abolished legislations that restricted and controlled the movements of Africans, and many moved to Lusaka in anticipation of a building boom in the capital of the newly independent state, working as unskilled labour in the building industry (Hansen 1997). Today, many of the descendants of the Mambwe migrants work as self-employed, unskilled craftsmen servicing the poor communities, but the majority engage in various forms of trade, which is a common survival strategy for the poor, urban majority in Zambia (Hansen 2004). Women and men travel to rural areas to purchase agricultural produce to trade in the towns, and villagers travel to the towns to trade their cash crops. The Mambwe and Mbala area are known for the cultivation of kidney beans, and descendants of the migrants engage in bean trade for the markets in Lusaka. The infrastructure of shops is poor in rural areas, and it is common to trade consumer goods in exchange for beans and other agriculture produce. Some of the Mambwe traders also travel abroad to neighbouring countries, in particular Tanzania, Zimbabwe, and South Africa, to purchase clothes, shoes, and consumer goods that they trade in Lusaka and the villages.

A characteristic of the Mambwe long-distant traders is that many of them keep up with relatives while travelling, a practice other chapters in this volume also discuss. The Mambwe trader's livelihood and mobility fully depend on the hospitality of relatives and kin in many different geographical locations. Relatives moor in each other's houses, which are dispersed across large spaces. Moreover, the networks of relatives and kin are important for the travelling trader's social security; if stranded, relatives may assist them in various ways.

The group of classificatory brothers settled in Lusaka knew some of the traders and travellers who came to stay in their houses very well; others they had never met. A particular characteristic of the mobile sociality and sociability among them, as among other ethno-linguistic groups in Zambia, is that relatives who hardly know one another – or initially do not know *about* one another – may stay in each other's houses for shorter or longer periods. The genealogical links between hosts and travellers may be remote, unknown, or there may not be any genealogical links at all. Strangers may trace genealogical links and possible other connections based on places of previous residence and work. Connections are crucial for the traveller's access to support and places to stay. What is peculiar with the present case compared to the aforementioned studies and conceptual framework is that kinship relations that in mobility research would readily be categorized as 'strong ties' may here be interpreted as novel and 'weak ties' that are construed among people on the move as strong and close ties through idioms of kinship. Interaction and exchanges in these novel and weak relations may be based on values and norms of generalized reciprocity characteristic of immediate family or permanent members of a household. Cultural and social practices of construing *relatedness* (Carsten 2000; Sahlins 2011a, 2011b) are thus in this case an important dimension of mobile sociality and sociability. I argue, however, that conceptualizing the forms of sociality among travelling Mambwe (e.g. staying in many different people's houses) as 'weak' and 'strong' ties may be fallacious. Kinship in this case is better seen as acts of social classifications (Lévi-Strauss 1963), and as falling within a particular category that *actualizes* a set of norms and obligations of reciprocity, mutual aid, and hospitality (see also Hays and Ninkova, and Engebrigtsen in this volume). Practices of recognizing some sort commonality or connectivity that enables inclusion in rather flexible categories are pivotal to Mambwe mobile sociality.

There has been a lapse in substantial ethnographic studies of kinship relations in urban Zambia. Studies were conducted in the 1950s and 1960s by Boswell (1969) and Epstein (1981) before kinship studies went out of fashion in anthropology and the later emergence of studies of kinship relations as cultural meaning and performance (Carsten 2000). A key concern for these pioneering studies of urban Zambia in the 1950s and 1960s was the flexibility of social categories on which an individual's network for social security and belonging was built. The studies showed that persons between whom genealogical links were remote or unknown built relations of reciprocity and mutual aid through idioms of kinship, parochial identities, or 'tribal' categories. As such, these pioneering studies prefigured the later perspectives on kinship as performance, which shows that kinship 'constituted in terms of procreation, filiation, or descent can also be made post-natally or performatively by culturally appropriate action and any combinations thereof' (Sahlins 2011a, 14). Sahlins (2011a, 2011b) argues for disentangling kinship from persons, and replace the former studies of kinship with a much more sociocentric perspective that focuses on transpersonal practices of coexistence and intersubjective being, which he terms 'mutuality of being'. 'Kinsmen are people who live each other's lives and die each other's deaths', he writes (2011a, 14), and continues,

> To the extent they lead common lives, they partake of each other's sufferings and joys, sharing one another's experiences even as they take responsibility for and feel the effects of each other's acts ... The constructed modes of kinship are like those predicated on birth precisely as they involve the transmission of life-capacities among persons. If love and nurture, giving food or partaking in it together, working together, living from the same land, mutual aid, sharing the fortunes of migration and residence, as well as adoption and marriage, are so many grounds of kinship, they all know with procreation the meaning of participating in one another's life. (14)

Descent and filiation structures much of the mobility among the Mambwe, but is not the only way of construing relatedness among them. In this chapter, I detail the ethnographic peculiarities of the social and cultural practices of relatedness of Mambwe-speaking persons on the move. Building on the previous studies of kinship and urbanization, I discuss the significance of the flexibility of kinship terms that enable people on the move to construe a sense of connectivity and mutuality of being and build relations of reciprocity and mutual aid. I begin my descriptions with a case of an unexpected visitor to the household in which I lived in John Howard Township, Lusaka, during my first field research in Zambia as a graduate student (1987–8),[1] which shows the dynamics and meanings of

1 The study is based in longitudinal fieldwork beginning in the late 1980s. I have followed up the initial research with numerous fieldworks and fieldtrips (1994–5,

kinship and mobility in the everyday life of the migrants. The incident took place on a sunny afternoon in January 1988, a few days after I had settled in.

The case of an unexpected visitor

I was sitting in the shade of a banana plant, chatting with my host. Suddenly, a young woman and a young child appeared just outside of the plot. The young woman held the child's hand, and carried a small suitcase in the other. My host went to greet her. He said, 'How did you leave the elders in the village?' 'They are in good health', she replied, and my host returned to his seat under the banana plant. The young woman went into the house and began to sweep and prepare for the evening meal.

I was puzzled by the brief exchange and that an apparent stranger would just enter the house and start doing household chores. Admittedly, my host had divorced a few weeks earlier; his ex-wife and their three-year-old son had returned to her parents in a village in Mbala district, but I did not expect him to remarry so soon, a thought that crossed my mind. Since his divorce, senior women among his relatives in the villages had begun sending him letters, 'recommending' different young women in the villages 'ready to marry' because they thought that he was in need of 'a woman in the house'. My host was fully employed, and, after his father died, he had looked after two of his younger brothers (aged sixteen and twelve) and a five-year-old son from an even earlier marriage. It was only later that I came to learn the moral imperative that a woman should engage in the housework wherever she is, in whatever house she finds herself. *Mama kutaya umwenyi*, a Mambwe saying goes; 'A woman is never a visitor'.

I asked my host to identify the woman, but he did not know her name or where she came from. 'She is a relative', he said. I wondered how he

1996, 2000, 2001, 2004, 2009, 2010, and 2015). Life history interviews combined with observations of residential patterns, frequencies of contact and exchange within the network at the given moments of time of the fieldworks have enabled me to observer changing patterns of mobility of individual members of the network over a generation.

knew she was a relative if he neither knew her name nor where she came from. 'I recognized her face', he replied. 'Oh, so you have seen her before?' I asked. After a few moments of reflection, he said he had seen her once, when she was a baby. He had recognized her because she resembled her mother so much, who was a daughter of his polygamist grandfather's fourth wife – she had last visited him in Lusaka six years ago. He now recalled the young woman as *nyina Richard*, 'the mother of Richard' – the boy she had brought with her. He knew the boy's name because relatives living in the villages often passed on messages through travelling villagers or sent a letter when any relative or kin there gave birth.

Rather inquisitively, I continued the conversation with him pondering the purpose for her visit to Lusaka, how long she would stay, and why she had chosen to come to this household since; after all, she was not of his clan and must have had clan members in Lusaka with whom she could stay. He surmised that senior relatives in the village worried about the well-being of his household had sent her to assist him. 'Anyway', he said, 'it is not in our tradition to ask such questions you are imposing now!' 'And don't you', he continued, 'become inquisitive and ask her these questions; she might think you act on my behalf!' He worried that word would spread among his relatives and kin that he was inhospitable and that they were not welcome in his house.

During the evening meal, two young men, the sons of one of his cousins living in the next neighbourhood, joined us, refusing the invitation to the evening meal. They came merely to sleep in the house, as they did every night. They always ate at their parent's house. The cousin had three sons, four daughters, and a young female relative who stayed as a dependent. Unless married, sexually mature persons of opposite sex cannot sleep in the same room, and the female cousin's house was not big enough for both her sons and daughters to sleep there. Neither did we have space for the mother of Richard. The house had only two rooms, both less than ten square meters, and we were never fewer than seven men and boys. After the meal, one of the younger brothers of the host escorted them to a classificatory brother (FFSS) who was living in an adjacent neighbourhood, and who had a larger house with bedrooms for both mature sons and daughters.

For the next week and a half, the mother of Richard came to our house early every morning and spent the entire day with us, keeping the house, looking after the two youngest boys, doing the laundry, and cooking for the family. She also planted a garden of maize on a vacant plot next to the house. She always left after the evening meal to sleep in the house of the host's classificatory brother.

It turned out that the mother of Richard knew nothing about the host's divorce. Rather, the elders in her village had advised her to look for him, because, as they had told her, 'he kept relatives well'. She came to Lusaka knowing only the name of the township in which he stayed. She had spent the first night with some friends at a neighbouring township. At the township market the following day, she had heard some people talking Mambwe, and she had inquired about the host from them. They happened to know who he was, and directed her to his house.

One morning, she did not turn up. My host worried about her absence now that she was under his care. In the evening, we received word that she had gone to a township on the other side of Lusaka to look for relatives there. After two days, she returned to collect her son and the luggage. She said that she would help a relative, her classificatory mother (MMZD), trading food, and that she therefore moved to the other township. The Mambwe practise sororate marriage and my host's polygamist grandfather married two sisters after his initial two wives had passed away. The younger of the two was her mother's mother. The older was a divorcee and had daughters from a previous marriage, one of which was the relative to which she moved.

The mother of Richard was one of many rural relatives who stayed in the house that year. A trader who regularly purchased goods in Zimbabwe for the markets in Mbala Town, the District centre of the village area often stopped over for one or two nights. A classificatory younger brother of the host was stranded in Lusaka for several months. He was a food trader between the village areas and Lusaka and a business partner had tricked him, and he had no business capital left. A young classificatory son of the host spent a couple of months there, after he had fled the village and hiding from the police who sought him for alleged cattle raiding. After the harvest in March and April, a few lineage members came to sell their agricultural produce at a food market near the township and spent the nights in the house. Others just came to visit for a week or two, but they were not many, as transport at that time was poor. It could take three to four days to reach town.

My host was a schoolteacher, and in between terms, he visited relatives in rural areas. He usually stayed in the village of the senior Mambwe chief where many of his fellow clansmen lived. His late father had a shop in the village, and moved there just a few years before he passed away. His birth mother and two of his birth sisters and their children still lived there. I accompanied my host on his visits, and while travelling to the villages we camped with his relatives and kin along the route. The oldest of his dependent younger brothers managed the household while he was away. Once when we were in the village, his five-year-old son moved voluntarily to a classificatory father in the township (FFBS) who was married. There, he had fewer responsibilities for household chores, and, as he said, 'I got breakfast'. The young boy's classificatory fathers left so early for school that they skipped breakfast. When we returned from the village, the young boy refused to come back.

On our second trip to the village, we also unexpectedly met my host's youngest son. After leaving Lusaka, his ex-wife had spent a few months in the village in which she grew up, but she needed to engage in long-distant trade to survive, which would be difficult with the custody of her little son. She had just travelled to the village where my host's mother and sisters resided, and left him with them since he was of their clan. In cases of divorce, it is common for children to stay with the divorcée until they are five or six years old, when, according to Mambwe conceptualization of child development, they are encouraged to become more independent of their birth mother. My host was of the opinion that it was too early for his son to leave his birth mother, but he could not be 'sent back' to her. Given the circumstances of the household in Lusaka, they decided it was best for the little boy to remain with his father's sisters in the village.

By accident, six months later, I met the mother of Richard at the food market in Mbala Town selling fritter buns. The business venture with her classificatory mother, trading dried fish from Lake Tanganyika to Lusaka, had failed. They had lost their entire business capital. Relatives living in the same township as her classificatory mother provided her with money for transport back to the village. There, she had left the child in the custody of her parents, and moved to relatives in Mbala in order to raise new business capital and re-join the fish trade. She apologized for not having visited before she returned to the village, but assured me that she would soon come to visit because she had almost accumulated enough capital. I did not see her again. It was a good harvest from the garden she had planted, and my host distributed all the maize to his relatives in the township as a token of their relationships.

Trajectories of mobility

The complexity of space and spatial mobility varies with the scaling of time (Hägerstrand 1970; Hannam, Sheller and Urry 2006). The mobility of a person during a day differs from the movements within a week, a month, a year, and a lifetime. Analysing the mobile history and co-residence of the different actors in the case above over a longer time-span sheds light on the flow of relatives and kin between households in the township, as well as on why my host had gained such a high reputation among kinsfolk that village elders would advise their fellows to seek his assistance while in Lusaka. The life trajectories of Mambwe informants show that their mobility and immobility is structured by different political and economic regimes of different scales and powers, such as the legislation and policies of the colonial and the post-colonial governments, regional and global systems of economic accumulation – and under the authorities of relatives and kin (see also Chapter 3 in this volume). In the case of the Mambwe, they are as individuals positioned differently in the relational field of kinship according to gender, age, and socio-economic strata; children and elders, for instance, may be moved between households and caretakers without much say on the matter.

My host belongs to a royal lineage and grew up in the village of the lineage chief, his paternal grandfather, together with a large number of classificatory brothers and sisters; descendants of both the previous chief and the present chief. When he reached adolescence, his father, who had established retail shops in several villages, sent him to the Copperbelt to purchase goods and ship them to the villages for restocking stores. While at the Copperbelt, he stayed with lineage members. He also completed his secondary education there. Due to his good results, the state educational authorities enrolled him at a teacher's training college in Solwezi, North-Western Province, where he stayed for three years. Upon graduation, he was posted in Luapula Province. As part of the nation-building policy, the independent government placed public servants in provinces other than those in which they grew up in order to combat nepotism and foster a common Zambian identity. While in Luapula, he was struck by disease. The state educational authorities offered him a position in Lusaka where he could receive better medical services. He asked to be transferred to the primary school in Chawama Township, where some of his agnatic brothers with whom he had grown up in the chief's village had moved. The school was short of staff houses, and he lived in the house of the oldest of his classificatory brothers (FBS) who had settled there, until the City Council allocated a plot for him in the adjacent township, John Howard, where he build the two-roomed house.

The oldest among the classificatory brothers settled in Chawama in the early 1970s. He quickly became a successful self-employed trader and broker for other lineage members' settling in Lusaka. He housed them, and they worked for him until they had secured their own jobs and built their own houses. Two of them had settled in John Howard, in proximity to the plot my host had been allocated; Houston, in whose house the unexpected visitor and her son slept, and Green to whose house my host's young son shifted when we journeyed to the village. However, we rarely saw Houston and Green. They both travelled in their work. Houston was a self-employed trader who bartered chickens from Southern Province with clothes on the Copperbelt. Green was a driver. When Green first arrived in Lusaka, his oldest classificatory brother in Chawama had sponsored his driving lessons. Car importers hired him on a temporary basis to drive cars from the ports in Dar es Salaam and Durban to Lusaka. Their wives engaged in mobile livelihoods as well to survive when the two brothers were on the road. Early in the morning, they left for farms outside of Lusaka to purchase vegetables, which they sold at markets down town. They returned to their houses just before the evening meal.

Houston had six children, four sons and two daughters. The two daughters were in their teens and had stopped school. They kept the house and looked after their younger siblings while their parents were away. Since my host had divorced, and there were no woman in the house to cook, as they said, my host and I regularly went to Houston's house for our meals. Green had two children; a boy and a girl aged eleven and five. The youngest did not attend school and was just left alone in the morning. When Houston stopped for a few days in the township on his travels between Southern Province and the Copperbelt, he often spent most of the day at the house of Green to look after the young daughter and to ensure that she had enough to eat. In some respects, the households of the three brothers in John Howard functioned as one. Every brother regarded the other brothers' children as their own, and all the children regarded the three brothers as fathers in whose houses they could sleep and eat.

In the mid-1980s, the business of the oldest brother in Chawama collapsed, and he survived on selling fritter buns at the local market in the township. He experienced this downward social mobility as troubling, and began to avoid some of his relatives. As word spread about his economic difficulties, relatives from rural areas became more reluctant to stay in his house while in Lusaka. Instead, they came to stay with my host who had a steady income and was in a much better position to house and feed relatives. This switch of roles strained the relationship between my host and his elder brother. Based on principles of seniority, the oldest man of any residential clustering of lineage members is supposed to be sort of a leader and provider for lineage members (Watson 1958). However, in Mambwe micro-politics in the villages and the chieftainships, brothers compete to take on this role. Such bickering among agnatic brothers is related to structural features within the Mambwe polity and the succession of chiefs and headmen, too complex to discuss here. Suffice to say that many Mambwe men put a strong value on being an overall provider and advisor for relatives and kin. This gives them a good reputation, respect from others, and self-esteem.

The mobile biography of my host and his classificatory brothers explains why the mother of Richard and other relatives and kin flocked to his house and the particular configuration of the co-operation between the households in the township, shouldering burdens of everyday life and of keeping visitors. The possibility to use my host's household as base for different activities, including my own field research, rested on a series of previous events, such as the chain migration of agnatic brothers to the township and their social mobility in town. The high frequency of contact and reciprocal exchanges between the classificatory brothers in Chawama and John Howard were partly based on descent, but also on a history of co-residence and spatial proximity. Up to secondary school, they had grown up in the same village, that of their grandfather who was the chief. The chief had thirteen sons altogether from his seven wives, and from these thirteen sons, thirty-four grandsons. The five classificatory brothers in Chawama and John Howard stemmed from four of the chiefs sons, and they knew almost all of their thirty-four classificatory brothers, but had no regular contact with them after they had left the village. The high frequency of contact and exchanges between the agnatic brothers in Lusaka was chiefly due to their residential clustering in Lusaka, and the geographical proximity of their houses. The face-to-face contact and exchanges were much more frequent among the three brothers in John Howard than between these three and the two brothers in Chawama a couple of kilometres away.

Common histories of co-residence and shared lived experiences are equally important, however, to descent and geographical proximity. One of the Mambwe migrant informants in John Howard lived nearby his brother born from the same mother and father. They had little or no contact. Coincidently, they had settled in the same neighbourhood. He had daily contact with a genealogical remote classificatory brother, however. There were no differences between the two birth brothers that could explain the lack of contact. When I inquired the one of them about it, he said that he and his birth brother had never lived in the same house, and they had hardly seen each other before. They had no need to keep each other's company, he said. He had, however, a history of co-residence with the classificatory brother he saw daily, previously to settling in John Howard.

Socio-economic strata also matter in the frequency of contact and exchanges between relatives. Another classificatory brother of the five brothers in Chawama and John Howard was a successful businessman and investor who owned businesses in the UK and South Africa. Only the oldest of the five had contact with the wealthy relative, partly because one of his sons was employed in one of the businesses. They never approached the wealthy relative for assistance.

My host's mobility through his life course was caused by circumstances beyond his control and under different authorities and regimes: the authority and economic activity of his father, the post-colonial state policy of enrolments at colleges and posting of civil servants, and personal misfortune. It was only later in life he was able to move where he desired. In the late 1990s, he recovered from his long illness and applied to the educational authorities for transfer to a school in his rural home area. There was shortage of teachers in rural areas, and his application was swiftly accepted. He put his house in Lusaka up for rent, and shifted to the rural areas. He did not settle in the village in which he grew up or where his birth mother and sisters resided, but in a village along the road between Mbala and Lusaka at the fringes of the territory of his lineage's chieftainship, where access to transport was much better. He also brought with him one of Huston's sons, born in the early 1990s, who had been given one of my names. The members of the lineage tend to accord me the same kinship position as my host. Most of his birth children and classificatory children address me as 'father' (*tata*), whereas others speak of me by using the kinship reference term with which they mention him. Since Houston had given his son my name, and had come to know me through my host, he thought it important, as he said, 'that his son stayed closer to his fathers' and asked my host to take him along. My host is now a pensioner, and one of the most senior members of his lineage. He still enjoys a good reputation among his relatives and kin, probably more than ever before. While out travelling, many stop to pay him a visit and seek his advice, others bring problems concerning the lineage as a whole, and classificatory sons and nephews have settled around him and assist him cultivating the land. He regularly goes to Lusaka for different purposes, but, due to differences, he does not keep up with his only classificatory brother that remains in the township. Contact and exchanges between them have discontinued. While visiting Lusaka today, my host usually keeps up in the house in Chawama in which he first lived when he was reposted from Luapula Province in the 1980s. His oldest classificatory brother passed away many years ago, and he stays with the daughter who now lives in the house, and whose obligation it is to care for ageing parents.

The frequency of visits and exchanges between the households of relatives and kin in Lusaka, and between relatives in rural and urban areas are as high, if not higher than in the 1980s. In the 1990s, when the HIV/AIDS epidemic was a major crisis for many migrant families, relatives mobilized to share the custody of orphaned children and shoulder funeral expenses. In the past decade, the Zambian national economy has had one of the highest annual growth rates in Africa, and massive import of cheap consumer goods, mostly from China (Carmody 2009, 2011), have escalated petty trade of consumer goods in rural areas, fuelling the circulation of relatives between rural and urban households. In the Mbala rural districts, this trade has taken the form of revolving markets, which increases the mobility of and frequency of contact between relatives scattered in different villages. Rapid economic development has also resulted in increased costs of living and social differentiation. Many of the poor Mambwe households are today, as then, not able to survive as autonomous households, but must rely on mutual aid between relatives and kin for social security.

The economic development has also resulted in higher competition and improvements in the transport sector. Today, there are several daily busses – of good standards and affordable prices – between Lusaka and the rural areas, which have increased the frequency and speed of mobility among the Mambwe. Nowadays, I meet in Lusaka senior members of the lineage with whom I stayed in the rural areas in 1988 and on return fieldworks in the 1990s. As elders, they stay with their children in town. In the villages, I meet Mambwe who lived in John Howard in 1987–8 and who have resettled in the rural areas.

Increased geographical and social mobility and better access to transport have also crated new opportunities for intermittent physical reunions. In the past few years, a new annual institution called 'family gatherings' (*ukulongana kwa lupwa*) has spread among the Mambwe in rural areas. The commoner clans and the royal lineages invite relatives from all over Zambia for a weekend feast in the village. The lineage to which my host and his classificatory brothers who settled in Lusaka after Independence belong had their first family reunion in 2013 in the village of the present chief. People think they now are so dispersed that they need to get together more often to know about one another. As my host and research assistant reiterated in many conversations with him over the years, discussing kinship relations, 'It helps to know one another', meaning that no one knows what the future brings and what kind of help one would need from relatives and friends.

Relatedness beyond Zambia

Not only Richard's mother, but also I was a surprise visitor to the household in John Howard in 1988. But, contrary to the Richard's mother, whom neither my host nor I have seen or heard from since then, I became more permanently entangled in relations of coexistence and exchange with the household members and their relatives. My host and head of the household was also my research assistant. A colleague at the University of Zambia (a remote relative of my host who had stayed in the household in the township during his own studies) suggested him as a research assistant. He wrote a brief note on a slip of paper as a letter of introduction, and advised me to contact his relative through the primary school in the township where he worked. Not before long, I had settled in the household. I paid my host wages for research assistance according to the University of Zambia regulations, but since I was a student, he would not charge me for lodging and board. Since then, I have been adopted as a member of my host's lineage and I always stay with them when in Zambia on field research, in both rural and urban areas. The relations established in the household in John Howard became paths for future mobility and engagements and, contrary to the case of the mother of Richard our relations grew. It seems as if I have adapted to a kind of sociality and sociability typical of travelling kinsfolk. They never charge me for lodging and board. But, from the moment I became permanently employed, they have expected that I contribute towards shouldering problems and concerns of household and lineage members. They expect from me what they would expect from any grown-up male lineage member in regular employment – contributions towards the education of the children, funeral expenses of lineage members, agricultural input such as hybrid seeds and fertilizers, business capital for women who engage in food trade, and travel expenses between rural and urban areas. They rarely ask for assistance outright; rather they raise problems and concerns, and are grateful for any contributions I can make. They are not exorbitant in their expectations. In rare cases, when people have bluntly asked for a large sum of money, others have immediately corrected them in public, saying that I am one of them, and that it is against their tradition to beg from and exploit relatives.

I am entangled in these relations while at home in Norway. They call me once in a while to inquire about the well-being of my Norwegian family and kin, and to report deaths and births. Occasionally, they also inform me about relatives in trouble moving me to condolences and the urge to make my contribution. What affects them, affects me. Thanks to mobile phones and swift money transfer, I can make my small contributions more frequently now than when we were reliant upon letters for communication and bank drafts. Mobile phones have also increased expectations with respect to the frequency of communication. For various reasons a few years ago, our telecommunication lapsed for many months, upon which I received a rather strong letter worrying about my welfare and indicating that such behaviour reflected poorly upon my character. When we communicate on mobile phones, we spend a lot of time exchanging information on relatives and recalling previous events. Given their patrilineal orientation, they never inquire about my wife or my late mother; their concern is the well-being of *mukombe* 'the old one' – my father, my siblings, and my children. They are in particular concerned with the well-being of my elder sister who came to see me in Zambia on my first field research, and of my oldest daughter, whom I was so lucky to bring along on field work in 2009. When they call me today, they are much more concerned with hearing about her than about me and about my youngest daughter, whom they have only seen in pictures.

Engaging in mutual assistance is but one situation in which I have been symbolically marked as a relative; rituals construe relatedness too. In life-cycle rituals, I must conform to the rules and norms appropriate to the ritual roles of the clan and the generation of my host. In a marriage celebration, I once danced with a woman I did not know was a member of the clan, which is against the rules. The Mambwe practise clan exogamy and one can only dance with persons of opposite sex from clans one can marry. For that evening and the following day, it was joked about that I had left them and was no longer of their lineage and clan.

Feasts make relatedness too. While in rural areas with my daughter in 2009, we kept up with my research assistant in the village along the Lusaka-Mbala road. We travelled to the village in which we had stayed in between terms in the 1988 and in which I had stayed on field trips in the 1990s, but, as usual, without any notice of our arrival. Also as usual, my host's birth mother and sisters sheltered us and fed us. They immediately prepared for a feast for my daughter, as they had for me at my first visit to the village in 1988, ritually marking my inclusion into their community. They brewed the ceremonial millet beer, *cipumu*, commonly prepared for the veneration of ancestor spirits (*uwengwa wa mipazi*). When the beer was ready for consumption, they summoned relatives, neighbours, and friends. As usual in these feasts, dances, pantomimic sketches and songs commence after an initially, solemnly consumption of beer. They performed numerous songs about their daughter – singing that she had finally 'crossed the mountains' to visit them, and for whom they had longed so much.

Kinship, urbanization and mutual aid: A view from earlier migration studies

The first studies of kinship and urbanization in Zambia was carried out in the 1950s and 1960s, in the then colonial Northern Rhodesia, by anthropologists affiliated with the Rhodes Livingston Institute (e.g. Boswell 1969, 1975; Epstein 1981; Mitchell 1969) and who were part of what came to be known as the Manchester School. In general, they focused on how migrants to the recently established mining towns formed relations and conceptualized the urban social environment, and how these relations and categories could be empirically researched and conceptualized in social theory. They pioneered the concept of 'social networks' and the anthropological studies of complex societies. Compared to villages where kinship organized political, economic and religious relations, urban kinship was primarily seen as a domestic domain. Relatives and kin were seen as links in an individual's personal network for social security in an urban situation, which the researchers characterized as being in flux – casual, fluid, unstable, and unpredictable (Boswell 1969; Epstein 1981). The vital thesis in these studies was that a sense of belonging, mutual aid, and reciprocity typical for the immediate family could be generated in relations built on idioms of kinship, but where genealogical relations were not known.

An individual's personal network that was seen as constituted on the basis of many different social categories, such as neighbours, work mates, fellow worshippers, friends, and relatives. Relatives in the network mobilized, co-operated and organized *ad hoc* to perform specific tasks, particularly in lifecycle crises such as at birth, marriages and funerals, and in personal crises such as unemployment, sickness, and litigation. Kinship relations came to symbolize social security; a safety net based on mutual confidence and reliance, where people took care of each other's interests (Boswell 1969, 1975; Epstein 1969, 1981). Kinship relations served instrumentally to address material needs such as financial support, housing, and work opportunities, and signified stability in a situation which otherwise could be experienced as unpredictable (Boswell 1969; Epstein 1981). In a sense, kinship relations expressed confidence and intimacy. Broadly speaking, kinship became synonymous with fundamental principles of equality and community and could be seen as part of the communitarian practices and values that Ferguson (1999) much later argued represented a genuine 'African-ness' in contrast to European exploitative colonial and industrial relations.

In the colonial towns in the 1950s, many urban migrants had none or few relatives residing in town and had to rely on a wider network of relations in which mutual aid could be generated. These extended networks included persons who people often referred to as relatives, and described within idioms of kinship, but with whom actual genealogical links were unknown (Epstein 1981). It became important in different social encounters to communicate kinship identities and trace possible kinship connections even if a possible connection did not change the instrumental content of the relation. Communicating sameness and connectedness could provide people with a sense of belonging. Frequent use of kinship terms as address terms in casual encounters was a way to perform a sense of belonging and explore prospects for assistance and future relations of mutual aid. Tracing possible kinship connections between strangers is as common today as then, and kinship terms are part of everyday talk.

The perspective on urban kinship as primarily a domestic domain and network for mutual aid was developed in contrast to kinship relations in rural villages (Epstein 1981). Kinship relations in small-scale rural societies were primarily seen as a public domain, providing for political, economic and religious organization and playing essential role in the formation of corporate groups. For many groups in Zambia, however, members of a clan typically lived widely dispersed and never operated on a day-to-day basis as a unified group or as an organized collective. The clan functioned primarily as a descent category. People who shared the same clan name claimed to be related, and common descent served as a category of interaction between strangers. If a person travelled outside his or her home area, it was likely that he or she was housed and fed by people who shared his or her clan's name. This was also extended to people from other ethno-linguistic groups who shared the same clan names and, hence, claimed to be related. The clan thus 'offered a symbol of solidarity and mutual aid to all those who could claim common decent' (Epstein 1981, 195).

In addition to belonging to a clan, people in Zambia reckon membership to a lineage of shallow depth – among the Mambwe, an agnatic lineage of three generations. The Mambwe differentiate between the clan of the chiefs and commoner clans. The clan of the chiefs are divided into several lineages, each with its own title and territory, comparable to a segmentary state (Southall 1988; Willis 1966). Both titles and territories have distinct names. Lineages and residential clusters of the commoner clans, of which there are more than fifty (Watson 1958), have no distinct names and are referred to as 'house' or 'hut' (*ng'anda*). Neither do members of a lineage or 'house' necessarily form a corporate group. In rural Mambwe societies, members of a lineage are widely dispersed and loosely organized. A Mambwe village is typically composed of members from many different clans and lineages.

Members of the same lineage in a village tend to group their houses and co-operate more with each other in production and consumption than with members of other lineages (Pottier 1988; Watson 1958). In this sense, there is no difference between urban and rural areas. The agnatic brothers in John Howard co-operated more with each other than they did with other agnates in Lusaka. Frequencies of interaction and exchange correlate to a certain extent with residential proximity (see also Chapter 5 in this volume). Frequency of contact and exchanges may increase among those who reside in a cluster. But, the residential clusters are not permanent or stable as people move, and their contact and exchanges may discontinue when they live apart.

In urban Zambia, relatives and kin have had fewer opportunities for residential clustering. Within the colonial, administrative town boundaries, residential areas for the migrants were initially either workers' compounds by the mines or servant quarters. However, in Lusaka in the 1950s, squatter settlements grew rapidly just outside the administrative town boundaries on non-arable land of European owned commercial farms (Hansen 1997; Williams 1986). These townships further developed rapidly after independence in 1964, when the independent government abolished the colonial legislations that restricted and controlled the movements of Africans. The independent government also expanded the town's administrative boundaries to include those settlements. Today, the majority of the population of Lusaka lives in these townships. Through processes of chain migration, there was initially a certain degree of residential clustering of relatives and ethnic affiliates in these squatter townships (Williams 1986), and, comparable to the villages, the clusters did not function as corporate communities expressing neat geographical or social boundaries.

Epstein (1981) argues that since descent in urban communities lost its function as a general organizational principle, the 'house' (*ng'anda*) in urban areas no longer denoted a lineage or a residential clustering of lineage members, and came to denote 'urban household'. However, one may question if the 'house' has lost its signifier as a descent category. The values of reciprocity and hospitality among Mambwe agnates that enables mobility spanning over large geographical areas are well captured in the rhetoric figure of a 'house'. In Chimambwe, *ng'anda* denotes a physical house and a lineage. The lineages can be seen as constituting a large, imagined house that comprises the physical houses of any agnate wherever the physical houses are geographically located. Anyone who belongs to the larger metaphorical house has rights to be fed and sheltered in each of the physical houses. These are the houses in which people can moor for longer or shorter periods, for whatever reason and independent of whatever form of mobility or mobile livelihood they are engaged in. It is to the large, metaphorical house people belong. In Chimambwe, there is no word for 'home' as such (Simonsen 2010). In contexts in which English speakers would use 'home', such as 'I am going home', they would say, 'I am going to the house' (*napita ku ng'anda*) or, 'I am going to the village' (*napita ku muzi*). They also speak about their clans and lineages as 'people of the same source' (*umwiko*). However, house, village or source does not necessarily refer to a particular place. Rather, they signify wherever relatives, with whom they would like to stay, live. In the late 1990s, due to rising living costs, several of the Mambwe who migrated to Chawama and John Howard in the 1970s and 1980s resettled in the rural areas. Like my host, many of them did not settle in the villages and places where they were born and raised, or from which they initially left when they migrated to Lusaka, but in villages where they found lineage members with whom they had good rapport and from whom they could receive assistance in accessing land and building houses.

Flexibility of kinship terms and patterns of mobility

The nature and boundaries of the social categories on which the individual migrant's personal network was based, was a central theme for the pioneering studies of urban Zambia. In his seminal paper on the Kalela Dance in the Copperbelt towns, Mitchell (1959) argued that 'tribal' categories were primarily an interaction category in casual encounters. The number of ethno-linguistic groups was too many for their names to function as social categories in the urban situation, and was reduced to a manageable set. People from smaller groups were subsumed under the category of the demographic and political dominant group in the larger geographical area from which they came. In Lusaka today, the smaller ethno-linguistic groups along the border with Tanzania, such as the Mambwe, Lungu, Namwanga, and Iwa are usually lumped under the category of Bemba. Many migrants from the four smaller groups settled in John Howard, and the Mambwe constituted the majority among them. Consequently, the Lungu, Namwanga and Iwa were subsumed under the category of the Mambwe. As interaction categories, people also included people from the same village areas, but who belonged to clans of other ethno-linguistic groups. For the traveller, the houses of people who can be defined as being from the same ethno-linguistic group or village area are also a potential destination and location for lodging and boarding.

The flexibility and inclusiveness of kinship categories are based on other mechanisms, characteristic of classificatory systems of kinship terms. In Zambia, they all belong to the classificatory variety that terminologically merges collateral with lineal relatives. In the Mambwe terminological system, ego's father's brothers are addressed with the same term and placed within the same category as the birth father, *tata* (plural and respect form, *yatata*). One's mother's sisters are addressed with the same term and placed within the same category as the birth mother (*mayo*, plural and respect form, *yamayo*). A man addresses his brother's children as 'my child' (*mwana*), and a woman addresses all her sister's children as such.[2] Mambwe migrants usually trace descent two generations back – to grandparents they have met – and the number of persons they address parents and sons and daughters increases substantially by employing the same-sex sibling link in each generation (F = FB = FFBS/M = MZ = MMZD; C = FBC = MZC = FFBSC = MMZDC). Life-story narratives show that many of the migrants in John Howard had grown up in many different households of their classificatory parents, and, thus, have history of co-residence with many of their classificatory parents and siblings. Examples are my host who during adolescence lived in the households of his classificatory fathers on the Copperbelt, his son who was left with his agnates in the village, and the man

2 In Chimambwe, there are no terms for 'son' and 'daughter' as such and discrimination is made with reference to sex (female: *anaci*/male: *monsi*).

who had not grown up with his birth brother. Many of the households of the Mambwe migrants in the township keep, for various reasons, classificatory children as dependents.[3] All adults have thus many children towards whom they have potentially parental responsibilities, and all persons have many parents who are potential obliged to provide them with food and shelter. Previous history of co-residence is important for the knowledge about the parents and, consequently, significant in the travellers and traders' decision as to where to look for shelter and opportunities for trade.

Relatives of the same generation, that is siblings and parallel cousins, also have obligations to look after each other. Based on same-sex sibling link in the above two generations (B = FBS = MZS = FFBSS = MMZDS/Z = FBD = MZD = FFBSD = MMZDD), these may constitute a potential large number of persons in an individual's social network. Norms of custody and care among them are based on gender and principles of seniority, and structure and pattern to certain extent the mobility of individual persons. The older cares for the younger. Men are expected to help the younger in the patriline of his generation, such as the elder brother of the host in Chawama assisted his agnatic brothers to settle in Lusaka. Before they marry, young women are expected to assist their older sisters keeping house and rearing children.

In Chimambwe, there are no terms for 'brother' and 'sister' as such. Siblings and parallel cousins are termed *ainane* as distinct from cross cousins, which are termed *avyala* (sing. *muvyala*). A person would address sisters, brothers, and parallel cousins with reference to relative age. A person addresses and refers to younger siblings and parallel cousins as *muto wane*, literally 'my younger one'. In conversations about persons in the same generation, referred to as third persons, further specifications are

3 Of fifty Mambwe migrant households in John Howard I surveyed during the field research in 1987–8, twenty-four kept dependents. Twenty-two of the fifty families had children living elsewhere, from former marriages or children who had married and established their own household. Altogether, fifty children lived elsewhere. Fifteen of them lived in rural areas with relatives. Most of these were sent to the rural areas to attend school because of poor options for enrolment in township in the 1980s. The rest lived with and helped their grandparents cultivating the land.

done with reference to sex (*anaci*/*monsi*) or by adding adjectives for age (*kalamba*, 'old [big]'/*nono*, 'young [small]'). A person addresses his older brothers and parallel cousins as *yatatalenzi*, literary 'father older than me or simply just *tata*, 'father' or *yatata*, 'father' in plural and respect form. This denotes that a man has parental responsibilities for younger siblings and parallel cousins, particularly to those of same agnatic lineage such as my host who looked after his younger brothers. When his polygamist father died in the early 1980s, he had five wives and twenty-seven children, of whom eleven had married and established their own households. The custody of the other children were divided between one of the deceased's younger classificatory brothers who succeeded the youngest widow, and the deceased's elder sons who took custody of the children of the other widows who had reached menopause, and hence, did not remarry through positional succession.[4] My host looked after the youngest of his brothers from his birth mother, and a son from his father's fourth wife.

A corresponding address term for a 'mother older than me' does not exist.[5] In principle, elder sisters have no comparable responsibility to take on the custody of agnatic siblings and parallel cousins. It is common, however, that a woman keeps a younger sister as a dependent in her household as a 'helper' (*kazwa*), looking after the children and doing household chores.[6]

4 Among the Mambwe intestate succession and inheritance (*ukupyana*) are patrilineal and positional (Watson 1958). A deceased man is most typically succeeded by a classificatory younger brother, who succeeds in the deceased's position, taking over his name, kinship relations, uxorial and genetricial rights, and property. Correspondingly, a deceased woman may be succeeded by a younger sister.

5 An elder sister is addressed *mangulenzi* (plural and respect form, *yamangulenzi*). The word *mangu* is difficult to translate but are used to address women to whom a person is related, but who bears children of other clans, such as 'father's sister' (*yamangusenje*), and 'mother-in-law' and 'daughter-in-law' who are addressed with the same term, *yamanguvyala*. As with the term for cross cousins, the suffix *vyala* discriminates relatives from agnates. The suffix *lenzi* denotes a person older than the speaker. Both suffixes connote person one should accord particular etiquettes of respect (*umucinzi*) manifest in rituals of greeting and parting, in conversational forms, and the type of topics they can discuss.

6 Younger sister of the husband may also be used as a 'helper'. A case in point is my host. After he remarried, he also kept a younger sister who was of marriageable age, but

The relationship is modelled after sororate marriage without the husband assuming marital rights. This practice is one of the chief reasons for the mobility of young, unmarried Mambwe women, and their migration to town. Another reason for the mobility of young, unmarried women is a particular role of the father's sister. A woman assumes a father's authority towards her agnatic nieces (BD). She can 'talk frank' to them and reprimand them. The migrants to John Howard sent rebellious teenage daughters to a person that they could classify as the girls' father's sister 'to be disciplined'.

Significant for the micro-mobility of boys and young men in the township is their relationship to men they can classify as mother's brother. They visit and keep company with them to escape the authorities of their fathers. A case in point is the five-year-old son of my host. When he matured in the mid-1990s and wanted greater autonomy, he often spent the afternoons and the weekends with a classificatory mother's brother down the road. Structurally, this relationship is opposite of the relationship between a father's sister and brother's daughter. The address term for mother's brother is *yayama*, which translates as 'male mother'. The mother's brother is a friend, and conversation partner; a person one can be open towards with sorrows and pains.

Conclusion

The Mambwe kinship system is a dynamic and inclusive form of social classification, a set of norms and values of reciprocity and hospitality, and a cultural orientation that enables and shapes the mobility of persons across vast geographical spaces, as well as the micro-mobility in their everyday life (for similar observations, see Chapters 1, 3 and 5 in this volume). Mambwe of common descent have clearly defined rights and obligations towards each other for custody and care. Every physical house of members of a lineage

who had chosen not to marry. My host's wife was also a schoolteacher, and the adult younger sister kept the house and looked after the children while she was at work.

constitutes a material infrastructure – fixed points in geographical space, which for the mobile person represents a potential destination and shelter for longer or shorter periods. This material infrastructure is not entirely fixed or immobile. Its range and geographical locations change as people shifts houses, and varies with the individual traveller's knowledge about them. The frequency of movement and exchange between the houses correlates with geographical proximity, for example, between lineage members clustered together residentially. In this regard, there are no differences between the rural and urban areas.

Identifying some sort of connectivity, such as possible links of descent and filiation, common acquaintances, or former places of co-residencies and work, is an important dimension of the mobile sociality and sociability of the Mambwe. The earlier migration studies' focus on the flexibility of social categories on which social networks of individual persons were built made sense in the context of mobility between relatives. However, their point of departure was the developmental and social consequences of circular labour migration in both rural and urban areas. They regarded town and country as two distinct socio-cultural systems and migration as a historical progression from a traditional rural society to a modern urban one. Moving to urban areas was seen initially as disruptive of social relations and institutions, and emphasis was placed on how migrants formed social relations and meaningful social environs in town as a matter of adaption to a different social system. In this sense, binary and sedentarist logics characterized their perspectives. However, treating town and country as two separate social and cultural systems quickly came under critical scrutiny. In the introduction to the proceedings from a conference in Lusaka in 1972 focusing on the relationships between town and country, Parkin (1975) writes what seems in phrasing and meaning identical to contemporary mobility literature:

> The volume [...] starts from the now generally accepted view that the rural and urban areas in East and Central Africa have to be regarded as part of one single field of relations made up of vast criss-crossing of peoples, ideas, and resources. (3)

Focusing on encounters with, adaption to, and survival in an urban environment, the multiple and intersecting forms of mobility between relative and kin discussed here fell out of the pioneering studies empirical focus and analytical gaze.

A study of mobility between relatives and kin needs a much broader concept than a concept of migration, as Piddington (1965) pointed out at the same time as the pioneering studies of kinship, urbanization and social networks in Lusaka and in the Copperbelt towns was published. 'Mobility' in a context of kinship refers to a wide range of contacts, interactions, and exchanges between relatives and kin clustered at different geographical locations (Hays and Ninkova in this volume). It covers everything from brief Sunday visits, exchanges and mutual aid among neighbouring kin, the sheltering of travelling kinfolk, holiday visits to the villages where urban migrants came from, and harbouring rural relatives who plan to settle in the city permanently, to indirect contacts and exchanges, such as the exchange of news, gifts, and money transfers. A concept of 'mobility' encompasses contacts and exchanges between relatives irrespective of forms of settlement and geographical distance, and, as such, differs from the concept of 'migration' in the pioneering studies, which presupposed permanent or semi-permanent settlement at the places of destination, or nomadic or seasonal patterns of migration between settlements (such as circular labour migration), which a concept of mobility does not.

A sense of connectivity and mutuality of being (Sahlins 2011a, 2011b) defeats time and space, and compassion, care, belonging, mutual aid, and generalized reciprocity are actualized and performed between them wherever they resides geographically. A compelling example of such mutuality of being is a surprise visit my host received when he was hospitalized in Luapula Province, where he was first posted as teacher. One day at the hospital, a senior woman and distant relative from Lusaka, visited him. She had travelled more than 800 km just to see him. He did not know about her, and he has never seen her since. The senior relative said that she was proud of him becoming a teacher, as that would help more of the children of their clan to achieve an education. When she had heard about his illness, she felt very bad and an urge to visit him fell upon her.

'Kinship accumulates or dissolves over time', Carsten writes (2013, 247) in a comment to Sahlins; there are 'processes of "thickening" and "thinning" of relatedness'. As the Mambwe moor in each other houses, their relationships and relatedness may develop and grow, wither away, or not develop at all. Relations of kin wax and wane over time and in phase with the life-cycle, as was the case of the relations between the five classificatory brothers. They grew up together in the same village, but lost contact when they as adolescents and young married men left the village and dispersed to different places in Zambia. However, their relations were renewed, and 'waxed' when they co-resided in Lusaka. Others form relations of kin as they move. Routes may become roots and pathways for the future.

The frequency of contact, exchange and micro-mobility among relatives and kin for most Mambwe correlates with residential clustering. However, analysing these relations through the metaphors of 'weak' and 'strong' or 'thin' and 'thick' *ties* – key to the studies of mobility, and kinship – is to smuggle in the backdoor a notion of space that may not be significant to them. The notion of 'tie' is a metaphor that we use, and we cannot assume *a priority* that other people conceptualize relations as 'ties'. I suggest that the Mambwe think more in terms of categories and systems of expectations, rights, and duties between categories. It seems they have a 'potential kin' and an 'actualized kin' when it is performed. Transforming that cultural notion into a 'tie' or 'bond' smuggles in a notion of space and potentially sedentarist epistemologies, which then lead to the 'strong tie' and 'weak tie' as functions of spatial distance. This is also present in our notion of 'social distance' when we actually mean varying degrees of mutual rights, duties and expectations between social positions (and position, again brings space in too). Can we think kinship 'non-spatially'? Or rather, do the Mambwe, in their performance of kinship, obliterate space?

Likewise, our notions of correlations between spatial and social distances are evident in the idea of 'thickening' and 'thinning' of kinship relations. This is like using a metaphor of a 'rubber band' rather than a 'tie' – the longer you stretch it, the weaker it becomes. The less stretched it is, the closer it is in spatial terms, the thicker it is.[7] Lévi-Strauss (1963) wrote that kinship is not about ties, but rather about a system of categories and

7 I am grateful to Lorenzo Cañás Bottos for stressing this point.

a set of obligations. Including a person in a particular category actualizes a set of duties, obligations, and expectations. A sense of relatedness and mutuality of being is among the mobile Mambwe not necessarily based on 'ties' or 'bonds', but on systems of social classifications, and obligations and expectations associated with those categories.

Bibliography

Bissell, David. 2013. 'Pointless Mobilities: Rethinking Proximity Through the Loops of Neighbourhood'. *Mobilities*, 8 (3): 349–67.

Boswell, David M. 1969. 'Personal Crises and the Mobilization of the Social Network'. In *Social Networks in Urban Situations. Analyses of Personal Relationships in Central African Towns*, edited by J. Clyde Mitchell, 245–96. Manchester: Manchester University Press, for the Institute of Social Research, University of Zambia.

——1975. 'Kinship, Friendship and the Concept of Social Network'. In *Urban Man in Southern Africa*, edited by Clive Kileff and Wade C. Pendleton. Gwelo [Rhodesia]: Mambo Press.

Carmody, Pádraig. 2009. 'An Asian-Driven Economic Recovery in Africa? The Zambian Case'. *World Development*, 37 (7): 1197–207.

——2011. *The New Scramble for Africa*. Cambridge: Polity Press.

Carsten, Janet (ed.). 2000. *Cultures of Relatedness: New Approaches to the Study of Kinship*. Cambridge: Cambridge University Press.

Carsten, Janet. 2013. 'What Kinship Does – and how'. *HAU: Journal of Ethnographic Theory*, 3 (2): 245–51.

Epstein, Arnold Leonard. 1969. 'The Network and Urban Social Organization'. In *Social Networks in Urban Situations. Analyses of Personal Relationships in Central African Towns*, edited by J. Clyde Mitchell, 245–96. Manchester: Manchester University Press, for the Institute of Social Research, University of Zambia.

——1981. *Urbanization and Kinship: The Domestic Domain on the Copperbelt of Zambia, 1950–6*. London: Academic Press.

Ferguson, James. 1999. *Expectations of Modernity: Myths and Meanings of Urban Life on the Zambian Copperbelt, Volume 57*. Berkeley and Los Angeles: University of California Press.

Glick Schiller, Nina, Tsypylma Darieva and Sandra Gruner-Domic. 2011. 'Defining Cosmopolitan Sociability in a Transnational Age. An introduction'. *Ethnic and Racial Studies*, 34 (3): 399–418.

Glick Schiller, Nina, and, Noel B. Salazar. 2013. 'Regimes of Mobility Across the Globe'. *Journal of Ethnic and Migration Studies*, 39 (2): 183–200.

Hägerstrand, Torsten. 1970. 'What About People in Regional Science?' *Papers in Regional Science*, 24 (1): 7–24.

Hannam, Kevin, Mimi Sheller and, John Urry. 2006. 'Editorial: Mobilities, Immobilities and Moorings'. *Mobilities*, 1 (1): 1–22.

Hansen, Karen Tranberg. 1997. *Keeping House in Lusaka*. New York: Columbia University Press.

——2004. 'Who Rules the Streets? The Politics of Vending Space in Lusaka'. In *Reconsidering Informality: Perspectives from Urban Africa*, edited by K. T. Hansen and M. Vaa, 62–80. Uppsala: Nordic Africa Institute.

Larsen, Jonas, Kay W. Axhausen and John Urry. 2006. 'Geographies of Social Networks: Meetings, Travel and Communications'. *Mobilities*, 1 (2): 261–83.

Lévi-Strauss, Claude. 1963. *Structural Anthropology*. New York: Basic Books.

Mitchell, James Clyde (ed.). 1969. *Social Networks in Urban Situations: Analyses of Personal Relationships in Central African Towns*. Manchester: Manchester University Press.

Mitchell, James Clyde. 1959. *The Kalela Dance: Aspects of Social Relationships Among Urban Africans in Northern Rhodesia*. Manchester: Published on behalf of the Rhodes-Livingstone Institute by the Manchester University Press.

Parkin, D. J. (ed.). 1975. *Town and Country in Central and Eastern Africa: Studies Presented and Discussed at the Twelfth International African Seminar, Lusaka, September 1972*. London: Oxford University Press on behalf of International African Institute.

Piddington, Ralph. 1965. *Kinship and Geographical Mobility*. Leiden: Brill

Pottier, Johan. 1988. *Migrants no More: Settlement and Survival in Mambwe Villages, Zambia*. Manchester: Manchester University Press.

Sahlins, Marshall. 2011a. 'What Kinship is (Part one)'. *Journal of the Royal Anthropological Institute*, 17 (1): 2–19.

——2011b. 'What Kinship is (Part two)'. *Journal of the Royal Anthropological Institute*, 17 (2): 227–42.

Simmel, G. 1950. 'The Stranger'. In *The Sociology of Georg Simmel*, edited by G. Simmel and K. H. Wolff, 402–8. New York: Free Press.

Simonsen, Jan Ketil. 2010. 'The Revival of the Mambwe-Lungu First-Fruits Ceremony in Zambia'. In *Reveries of Home: Nostalgia, Authenticity and the Performance of*

Place, edited by Solrun Williksen and Nigel Rapport, 45–66. Newcastle upon Tyne: Cambridge Scholars Publishing.

Southall, Aidan. 1988. 'The Segmentary State in Africa and Asia'. *Comparative Studies in Society and History*, 30 (1): 52–82.

Urry, John. 2002. 'Mobility and Proximity'. *Sociology*, 36 (2): 255–74.

Watson, William. 1958. *Tribal Cohesion in a Money Economy: A Study of the Mambwe People of Northern Rhodesia*. Manchester: Manchester University Press.

Williams, Geoffrey J. (ed.). 1986. *Lusaka and its Environs: A Geographical Study of a Planned Capital City in Tropical Africa, No. 9*. Lusaka: Zambia Geographical Association.

Willis, Roy G. 1966. *The Fipa and Related Peoples of South-West Tanzania and North-East Zambia*. London: International African Institute.

NIGEL RAPPORT

Epilogue: A Mobility Perspective and the Writing of Existential Power

Preamble

This epilogue has two aims. First to review how the 'mobility perspective' outlined by the book's editors manifested itself in this book. Second, to consider more broadly mobility as an anthropological way of knowing. The strength of the discipline, it can be contended, is not found (primarily) in theoretical abstraction but in the close reading of human lives. Anthropological insight is grounded in closely living with others: the human condition is approached in anthropology through an appreciation of individually embodied experience – both that of the anthropologist and that of research participants – and this experience concerns the imaginative and practical construction of worlds of meaning in particular times and places. Movement is key: key to the way in which individual embodiment is experienced and knowledge of the world is constituted; key, too, to the way in which social relations and cultural traditions come to be constituted. The expertise of anthropology is to witness the intrinsic connection between these: how the sociality of movement and the individuality of movement are versions of the same set of human experiences. It is from people in motion that human societies are made. A mobility perspective holds out the promise of bringing together the sociological and the phenomenological as intertwined.

A look back

> Je haïs les voyages et les explorateurs.
>
> —CLAUDE LÉVI-STRAUSS (1981, 1)

> Travel (something the ethnologist mistrusts to the point of 'hatred')
>
> —MARC AUGE (1995, 86)

The book's tenets

An appreciation of movement is intrinsic to an anthropological apprehension of human perception, of personal consciousness, and of social relations, I would argue. The moving body of the individual human being comes to know – sense, perceive, interpret – its own distinct things. Movement is intrinsic to the sense we make of self and of world, the identities and relations we construe. Human social life concerns how these individual, knowing bodies impact upon one another: how the knowledge each contains is deployed in interaction with others and is exchanged – however partially and distortively – in the constitution and maintenance of social relations and cultural conventions (Rapport 1993, 1997, 64–79; Rapport and Dawson 1998).

These are by no means novel claims. Gregory Bateson famously averred in the 1950s and before: 'all knowledge of external events is derived from the relationship between them from relations that the brain conceives as being *between* things' (1951, 173). To conceive of things and relations is to move those things relative to the point of perception (the brain) or relative to other things within the field of perception. Movement is fundamental to the setting up and the changing of relations by which things gain and maintain and continue to accrue identities. Indeed, since one of the 'things' that thus comes to possess an identifiable character is the individual perceiver, movement is fundamental to the identity of both personal selfhood and of the social relations that individuals construe.

There are a number of corollaries of this thesis, as Bateson elaborated. The first is that the identities and relations that emerge from movement are both material and immaterial. 'Mothers' and 'beauty', 'justice' and 'jewels', 'novels' and 'podcasts' are all outcomes of movement relative to a point of perception. Movement is essential to the practice by which an individual mind perceives and constructs an environment, whether 'natural' or 'cultural' or 'virtual' (Bateson 1980, 100ff.).

A second corollary is precisely that the mind is individual in this regard. Movement is relative to the individual perceiver. Bateson attends to this by describing the individual mind as 'an energy source' (1973, 126), responsible for energizing the movements that underlie the perception of difference and for perceived events in the world – the mind is not merely being impacted upon by environmental triggers. More generally, each human individual is an 'energy source' inasmuch as the energy of his or her acts and responses derives from their own metabolic processes – not from external stimuli. It is with this energy, through this movement, and by this construction of relations and objects, that individuals create order and impose it on the universe. 'Order' is a certain relationship between objects that an individual mind comes to see as normal and normative; it is dependent on the eye of the individual perceiver and may not be what others perceive as orderly. Human beings are active participants in their own universes: 'disorder' and 'order' are statements of relations between a purposive perceiving entity and some set of objects and events; they are determined by individuals' states of mind. A final corollary, then, is that in the interaction between two or more such individual 'centres of energy' – when human beings interact with one another in social exchange – 'schismogenesis' is a likely outcome (Bateson 1936). It is likely that the orderly environments – the version of self and world – of different individual interactants will themselves differ, and that these differences will eventuate in miscommunication of varying degrees between human beings who attempt to share their perceptions and interpretations with one another. Movement is foundational of what is perceived by individuals, and thus foundational of the social relations, social structures and cultural traditions in turn constituted and maintained by those moving bodies; the consequences of movement are fundamental to the nature of social interaction – schismogenetic, distortive – as they are to personal consciousness.

While these ideas are not new, I would nevertheless concur with Paul Carter's (1992) call for a new 'migrant perspectivism':

> It becomes ever more urgent to develop a framework of thinking that makes the migrant central, not ancillary, to historical process. ... An authentically migrant perspective ... might begin by regarding movement, not as an awkward interval between fixed points of departure and arrival, but as a mode of being in the world. (7)

The continuing tendency in anthropology (and ethnology), as Marc Auge's epigraph to this epilogue suggests, has been to conceptualize a normative relationship between identity and fixity. Anthropological analysis has sought to connect identity with specific times and places – at the very least with spaces – so that to be at home in the world was to be environmentally fixed: if not stationary then at least existentially centred in a home terrain.

What is missing is an appreciation of movement as home in itself: the way in which home, identity, relations are themselves manifestations of (versions of) human movement – home *as* movement. As Sørensen and Olwig adjure (2003), it is necessary to take movement as the point of analytical departure: movement not as disruptions to individual lives and as breaks with normativity but as intrinsic parts of human life trajectories. Auge's claim is no doubt a caricature but it is also indicative: Lévi-Strauss (1975, 14–30), for instance, seeking to characterize myths as mechanisms for the suppression of the sense of passing time and space, and enshrining fixed points from which the world took and takes shape. Myths then rooted cultures in time and space, as cultures rooted societies and their members in genealogies of 'blood, property and frontiers' (Carter 1992, 8).

This assumption has been increasingly called into question: we have witnessed a shift in the norms of anthropological commentary. As James Clifford summed up, the world can no longer be sufficiently construed in anthropology as divided up into framed units, territorial segments and the like, each of which shares a distinctive, exclusive culture. Rather, we find ways of life 'increasingly influencing, dominating, parodying, translating and subverting one another' (Clifford 1986, 22). The notion of fixed and bounded cultural worlds gives way to a recognition that *all is situated and all is moving*: 'everyone is now caught between local origins and a cosmopolitan society in which all humanity increasingly participates' (Hart

1990). Movement, according to John Berger (1984, 55), is the 'quintessential' human experience: emigration, banishment, exile, labour migrancy, tourism, urbanization, and counter-urbanization are the central motifs of modernity. Equally, anthropological commentary has sought to redress notions of time, not now treated as a unit of measurement by which social structures and cultural traditions can be found to reproduce themselves (Rapport 2014, 416–21). Time is now a means to recognize human life as a constant becoming: the individual moves through moments of consciousness and through phases in a life-project; social arrangements and cultural conventions that have longevity are nevertheless being constantly reworked, both by their members and by their wider environments (Nielsen and Dalsgaard 2015, Nielsen forthcoming).

To highlight movement (and flux) in this way is not to essentialize it, moreover: to claim 'it' is somehow always the same movement. Human movement remains a polythetic category of experience, diverse, and without common denomination in its particular manifestations. Nor is it to underrate either the forces giving rise to large-scale physical movements of population in pre-modernity (famine, plague, crusade, imperial conquest, urbanization, industrialization), or the forces still arrayed against such movement (restrictive or repressive state or community institutions, state or community borders). To talk about the quintessential experience of movement is neither to deny power and authority nor the differential motivations and gratifications in that experience that uses of power might give onto. Rather, what is drawn attention to is *the human nature of movement*. Movement is at the centre of how we, as human beings – migrants and indigenes, tourists and locals, refugees and citizens, urbanites and ruralites – construct experience and identity, personally and socially. And hence, the multiplication and diversity of analytical terms to which anthropology, in the wake of its recent recognition of the ubiquity of movement, has had recourse, including: globalization, creolization, hybridization, synchronicity, syncretism, cosmopolitanism, and the 'flux of becoming'.

Disagreement remains over whether the anthropological thesis that now links movement and identity pertains to a *historical* phenomenon or an *existential* one. Is it a matter of authentically *representing* an abiding human truth or is there something essentially *contemporary* about this movement? The historical argument is the harder one to sustain, remaining prone to the grand-historical reductions of earlier generations of anthropological theorizing: from 'fixity to movement' as from 'mechanical solidarity to organic solidarity' or 'community to association' or 'concrete thought to abstract thought' or 'holism to individualism'. Certainly, Bateson's propositions with which we began presume a universal relevancy, while the history of human movement makes generalizations about the uniqueness of the present short-sighted.

In their editorial introduction to the present volume, Kjersti Larsen, Jan Ketil Simonsen and Ada Engebrigtsen set out seven clear tenets concerning how mobility is to be treated:

- The collection should focus on mobility in social life, especially on oscillations between movement and fixity in certain social spaces, as construed by anthropologists who have carried out long-term fieldwork in their areas of study.
- The empirical material on which the discussions are based need not, however, initially have been collected with the theme of mobility in mind, nor among people traditionally deemed mobile. The analytical appositeness of a mobility perspective for anthropological thought and theorization may thereby be tested.
- The ethnographies should concern life histories, actors in time: individuals as well as households, the evolution of personalities, families, communities and networks.
- Different time-scales may be employed, so that the movement and connectivity of persons and groups within the life span of an individual, a family, a household, or even a generation, may be compared against the *longue durée*.
- The ethnographies should engage with the consequences of socio-political and economic change, in particular the way that national and international factors impact upon values, ideas and practices in local life-worlds.
- The ethnographies should consider the relations between spatial or physical mobility and social or economic mobility over time.

- Mobility should not be seen as necessarily indicative of social disorder and cultural disruption – or not in a negative way – but as possibly constitutive of social and cultural process as such.

Perhaps the key question posed by the editors concerned how mobility was experienced, and how it was different from what was not mobile. 'How do those who have been physically (and perhaps dramatically) "on the move" in their lives experience and perceive "belonging" and "connectivity"?' 'How does an experience of movement fit in broader life-worlds and life-projects?' It is, above all, a *local* understanding that the volume pursues: a so-called 'view from below' as opposed to the supervenient view that might presume a generalized knowledge of *what* 'mobility' is, and *who* is mobile, and *how* a local life is affected mobility-wise by exterior factors and forces. In Anthony Cohen's apt summation: 'If we do not do descriptive justice to individuals, it is hard to see how we could do it for societies' (1992, 229).

The book's contributions

How have these concerns been made manifest?

Øivind Fuglerud considered that in public discourse and to some extent in academic literature, there has been a tendency to regard the processes of societal integration and transnational engagement as forces that pull in opposite directions. For Tamil refugee-migrants in Norway, however, a transnational orientation towards a Tamil 'homeland' (Eelam) has over time helped to overcome what has been perceived as exclusion from the Norwegian host society and has also promoted social mobility within Norway. Refugee-migrant Tamils in Norway have been able to forge a 'horizontal' solidarity within the Tamil community that is anchored in a national identity, which to an extent overcomes traditional hierarchical, caste differentiations. That is, over the last three decades in Norway a class-based, internal solidarity, at least between members of the upper levels of the Tamil community has been developed, based on a co-ordination of interests, which has in turn assisted integration into Norwegian society. New Tamil networks now connect people of formerly different families, villages, and castes. These networks at once look back to an ethnic origin and forward to new national integrations. A perception of global movement is intrinsic to a sense of national rootedness.

For Ada Engebrigtsen, both the Norwegian Roma and the Romanian Romanies in Norway are in a position where their cultural norms as mobile subjects are more or less 'illegitimate' and seen as a threat by Norwegian authorities. The norms cannot be presented as part of claims to identity in public negotiations with the wider Norwegian society; publicly, the Roma must be seen to 'play along' while exploiting 'cracks in the state's armour' for the purpose of maintaining traditional mobile practices. Physical mobility is vital for a 'nomadic' Roma identity – which also renders these populations oppositional to state power and control that call on ethnic groups to possess territorial roots in order also to possess legitimate interests and claims, and to develop legitimate strategies of negotiation with the Norwegian majority. Tactics of accommodation to norms of fixity on the one hand and differentiation from these on the other are hence deployed alongside one another in order to be tolerated in Norway while at the same time retaining traditional livelihoods and community belongings that are seen to derive from perceptions of independence and freedom to move. The Roma are under constant surveillance as part of a Norwegian regime of pastoral care and are either subjected to programmes for cultural change or the politics of rejection. Roma practices of continuing mobility enable majoritarian and state power to be continually parried, even if never obviated or conquered.

Kjersti Larsen explained how a focus on movement and having supra-local, national and transnational connections provides a certain social cache and a higher level of social standing in Zanzibar Town. Historically, complex itineraries across the Indian Ocean over extensive periods of time were instrumental in the forging of new socialities and new societies. From the 1960s, however, the newly independent nation-state of Zanzibar, especially in its post-revolutionary guise, aimed to limit the forms of mobility that had been previously so formative, and insisted on clear state jurisdiction over specific territories. Notwithstanding, even while openings and opportunities for spatial mobility – and for social mobility – have become more limited for most Zanzibaris than they once were, local discourse still involves spatial movement as essentially nourishing of future prospects.

People of all age groups continue to engage in discussions about travelling and migration, and about places and lives elsewhere. In short, perceptions – and also practices – of movement, of trans-local connections and having had experience of life in other places, remain integral aspects of livelihoods both past and present. Persons, families, and households remain connected transnationally, and these connections continue to imbue modes of social mobility within Zanzibari society.

In Jennifer Hays' and Velina Ninkova's account, among the Ju|'hoansi San of Namibia, everything is always in motion: there is no such thing as stasis, only different tempos and patterns of movement. There is *neither* 'vertical' mobility in social status *nor* horizontal mobility in space. Again, for the Ju|'hoansi, these variables intersect in complex ways. Much of the discourse of socio-economic development among governments and NGOs plan to increase food security and overall quality of life for the Ju|'hoansi assume an idealized vertical progression in local life. However, only by recognizing the significance of a horizontal axis – the way in which social relations, identities and loyalties pertain to relationships to land – *and* the legitimacy and efficacy of horizontal mobility as a risk-management strategy, can any approach successfully take into account local perceptions and understandings.

Arne Aleksej Perminow considered how a Tongan sociability based on the social significance of detachment, reattachment, mutual flows, and nurturing spaces may be threatened in the long term, as more people move overseas and fewer remain to uphold local practices. The argument can be made, nevertheless, that the socially constitutive force of cultural notions about how relations are established and maintained through 'nurturing, feeding or caring for the space between' (*tauhi vaha'a*) remains considerable *wherever* Tongans may be based – and for however long. The cultural notions and their capacity to establish and maintain social relations may traditionally have found their 'natural' setting and have been most easily be observed in the movements, flows, and changes of place of substances, people, and things in everyday village life. However, the cultural assumptions and their socially constitutive force have not been left behind when the village itself was left and nor when the Tongan border is crossed. The norms of movement and flow in nurturing spaces continue to imbue Tongan life with a significant and differentiating ethos wherever congregations might be established.

Finally, Jan Ketil Simonsen described the kinship system of the Mambwe in Zambia as a form of social classification and a cultural orientation that is both dynamic and inclusive and that enables (and shapes) the mobility of persons across large geographical areas. The dynamism of the system means that there are both 'potential kin' and 'actualized kin', but Mambwe who can claim common descent have clearly defined rights and obligations towards one another for custody and care, compassion, and mutual aid. The sense of connectivity and mutuality of being that kinship affords overcomes time and space and ensures a generalized reciprocity that is actualized and performed. It was in the mid-1960s that the Mambwe first engaged in a process of chain migration, settling in the capital Lusaka more than 1,000 kilometres away from their home villages. Today, while many of the descendants of those first migrants work as self-employed, unskilled craftsmen servicing the poorer communities, most engage in forms of trade. Women and men travel to rural areas to purchase agricultural produce for trading in towns while villagers still travel to the towns to trade their cash crops. A characteristic of these long-distant travels is that the mobility depends on the hospitality of kinsmen at different geographical locations. Relatives in transit 'moor' in each other's houses for longer or shorter periods – whatever the state of their prior knowledge of one another. The genealogical links between hosts and travellers may be remote; they might indeed be previously unknown; host and traveller may even give up attempting to specify the genealogical links. What is key is the flexible way in which kinship is deployed: the dynamic system makes ostensibly strong ties perceived in practice as unnecessary and ostensibly weak ties, construed among people on the move, as strong and close, capable of instilling a perception of mutuality of being and of building relations of reciprocity and mutual aid. Every physical house of members of a lineage constitutes a node in a network in geographical space that represents a potential destination and shelter for longer or shorter periods for Mambwe individuals on the move.

As a whole, the book's contributions point to the ubiquity of movement in these human lives, and the ways and extents to which lives are lived *through* movement. A diversity of kinds of movement is described, physical, social, economic, virtual and imagined, esteemed and hidden. Order and sense (meaning) are accrued through movement. Whatever the forces and factors militating against movement – seeking to control movement in the name of a more static and fixed ordering of the world – movement is maintained notwithstanding. Time measures kinds of flux.

A look forward

> When you see what people actually do, as distinct from what they are supposed to, most of the category distinctions which provide to an orderly framework for our social thinking tend to disappear.
>
> —EDMUND LEACH (1977, 11)

> It's not always where and when you were born that matters; it's where you were reborn ... Our humanity is only fully realized when we suffer the ordeal of a second birth.
>
> —MICHAEL JACKSON (2015, 16, 26)

Anthropology's tenets

Michael Jackson describes a fundamental aspect of the human existential condition. We practise a capacity to create identities and world-views for ourselves. These pertain less to the accident of birth – of geographical and sociocultural location, even of how others view us – than to how we come to view ourselves relative to what we interpret to be our sociocultural and natural environs. It is a kind of second birth that we undergo, or keep on undergoing, as we make sense of ourselves throughout our lives; continual moments of 'rebirth'. Jackson's description also resonates with sentiments of Edmund Leach, who describes human beings, 'essentially', as 'criminals by instinct' (1977, 19). He means that relative to the orders, statuses, and roles we might be surrounded by at birth – the world of social and cultural categories in which we are ascribed – the individual identities we instinctually inhabit are necessarily and naturally a-structural and anti-structural. Our nature – and our behaviour – is non-systemic relative to these environing orders and norms for we can but be our unique selves. And finally, there are significant resonances between Jackson and Leach and the case studies provided by this volume. To admit a mobility perspective into ethnography and analysis is to become sensitized to individual human beings who deploy the common sociocultural categories of their birth according to personal perceptions and according to experience that is grounded in movement and moment. The ground of our anthropology is the different consciousness or world-views inhabited by different individual human actors at different moments of their lives, a consciousness that owes its character and its contents to the movements each actor causes himself or herself to make. Here is the human being as a kind of projectile, set on a life course of individual movement, energized by meanings and intentions – by 'life-projects' – developed in sociocultural environments but deriving from his or her own irreducibly individual acts of perception (Rapport 2003). In Georg Simmel's terms (1971, 24), between the forms and structures of social life and individuals' vital 'impulses' there is a 'reciprocal' relation: human beings navigate among social-structural institutions and cultural conventions as means and modes of their livelihood, belonging, expression, and communication. However, individuals' 'native force' causes them to move according to their own logics and to accrue meanings and maintain senses of self and other particular to themselves. When sensitized to this the anthropologist discerns a radical break between the *apparent* order of social structure and symbolic system by which social and cultural life is ostensibly practised and the actual diversity – potentially chaotic – of how those structures and symbols are animated and 'inhabited' by individuals moving through the moments of their lives.

The contributions in this book urge us to consider a new beginning where anthropologists base their understandings not on the apparent fixity of the categories that seem to be imposed on social and cultural life (by virtue of social structures and symbolic systems) but on daily lives. When are people not 'in movement' between moments of consciousness? When do the cognitive and physical movements that people make not also concern their social and cultural identities? Where does movement through 'local' spaces stop and the 'supralocal' begin? When are the life histories of communities, networks, households and families not at the same time those of particular people? Of what is the *longue durée* composed if not the life spans of individuals? Ethnography discloses the 'view from below' – the essential transitoriness, momentariness and 'gratuitousness' of life individually experienced and socially exchanged (Rapport 2008) – and it is this nature and condition that should be treated in anthropological analysis.

Anthropology's contribution

> Life is a luminous halo, a semi-transparent envelope surrounding us from the beginning of consciousness to the end. Is it not the task ... to convey this varying, this unknown and uncircumscribed spirit, whatever aberration or complexity it may display, with as little mixture of the alien and external as possible? ... Let us record the atoms as they fall upon the mind in the order in which they fall, let us trace the pattern, however disconnected and incoherent in appearance, which each sight or incident scores upon the consciousness. Let us not take it for granted that life exists more fully in what is commonly thought big than in what is commonly thought small.

This is Virginia Woolf in 1938, imagining how one might write literature in such a way that life, as she put it, did not 'escape' the text (1938, 148–9). Woolf sought to initiate a fictional paradigm sensitive to the stream of human consciousness – for which she herself and the likes of William Golding, Stevie Smith, and James Joyce were to become renowned exponents. The discussion has a continuing relevance for anthropology too, I would argue (Rapport 1994), echoing debates we have had and continue to have on 'life-writing' that sees beyond the surface of conventional forms. Michael Jackson has described the work needed to annul the language of 'cultural essence', then, to 'purge our discourse of the idealist connotation of the culture concept' that causes us to 'elide the line that separates words and worlds, language and life' (Jackson 2002, 109, 115). Woolf recognized the way in which privileging the normative categories of social life could lead to life as it was individually led and experienced 'escaping' the page; Jackson, similarly, would draw anthropological attention to how 'culture' has come to be treated in a fundamentalistic or essentialistic way as a thing-in-itself: not as 'an idiom or vehicle of intersubjective life' but as 'its foundation or final cause' (Jackson 2002, 125). We see the fixity of categories – not the flux of life. The empirical reality of the human condition consists in the tension and the interplay between persons and categories: to privilege the formalism, finitude and fixedness of the latter is to reduce and traduce the 'open-endedness and ambiguity of lived experiences' (Jackson 2002, 125).

How precisely might anthropology contribute to this phenomenological effort in ways that are appropriate to it being a social science? What might a piece of anthropological accounting offer, in the spirit of Woolf's project of capturing 'life' that connects the phenomenological and the sociological? I end by offering an ethnographic vignette. It pertains to a fieldwork as a hospital porter in a large public hospital in a city in eastern Scotland (Rapport, 2008); it is an attempt to recount what constitutes a person's life as it is as opposed to what the categories and conventions of their contingent social and cultural placements deem it to be.

ROGER AT CONSTANCE

Constance Hospital, Easterneuk, is a teaching hospital with a full range of medical specialisms and a working population of thousands, including some 130 porters. Not being medically trained, porters occupy a somewhat lowly position in the organization. Their tasks call for physical stamina more than any other skill, and their metier is movement: throughout the working day porters are responsible for ferrying patients across the large and wide hospital plant, for delivering mail within the hospital, and for transporting body parts, bodily substances, and dead bodies between different sections of the complex. Much social interaction is conducted on a hierarchical basis, however; the porters adapt to the needs of doctors and administrators, nurses, and clerical staff, whose specific skills attach them to specific sections of the hospital as a workplace, fixing them to a space, while the porters' official physical location is transitory.

The porters do, nevertheless, have a space of their own: their lodge or 'buckie', which consists of two cramped but warm rooms below the level of the main hospital concourse. Here, the porters are relatively stationary, resting between jobs, reading the newspaper, listening to the radio, making tea and sandwiches, eating, making jokes, sparring, fighting, and gossiping – until the telephone rings again and the portering chargehand sends one or more out once more into the hospital plant to fetch and carry. But the stationariness is relative. The porters stand and pace in their buckie; they are full of activity; most significantly, their talk moves them constantly to worlds of experience, imagination, intention and hope that dwarf their world of work at Constance. In their buckie, the porters keep up a commentary on life – the hospital, the wider Easterneuk world, one another – whose ethos is critical and humorous, even cynical. The commentary calls into question so-called claims to expert status, skill, and even identity. Being 'a man' (all but two of the porters are male) is surely an easily identifiable matter, and 'manliness' supervenes upon the conceits of academic medicine with its fetish for rank, class, and ritual. In other words, Constance Hospital's categorial and normative world of work becomes insignificant in the context of a *man's* life of fun, 'crack', joking, drinking, fighting, sex, and football. Unlike many of the hospital staff, the porters mostly hail from the local area; hence, to be at work at Constance is an extension of local movements and relations to which they already and assuredly belong. The porters are at home in Easterneuk in a way that makes their time in the hospital secondary, almost irrelevant.

Roger Weir was a porter with whom I came to regularly share lunch breaks. He and I had begun as porters at Constance on the same day, and we had undergone the same induction procedure. Roger was some twenty years my junior at the time (in his early twenties) but we chose regularly to take our lunch-breaks together and to sit in the Staff Refectory and compare notes on our working mornings, amid other comparisons – the sandwiches we had brought and the recreations and loves we had beyond work. In particular, we would match the ways in which the hospital was endeavouring to inculcate us into its perceived institutionalism and hierarchy. Here was Roger and here was me, with our distinctive life-projects, our plans for progressing through our workdays and our lives, and here was the institutional world of the hospital: our conversations compared the ways and extents to which we were each accommodating these disparities and disjunctions in our lives.

Here is an extract from a sample lunchtime exchange:

ROGER: I spent £60 boozing Saturday and yesterday combined: not bad! It was 3.30 in the morning before I got to bed. I wasn't gonna go out, then my friends came round at 9.30 and convinced me. We had a great time! Head-banging away to the music!

NIGEL: How can you survive on so little sleep?

ROGER: I don't know. It amazes me too! But the tiredness is something I can put up with at work: it's hangovers that get to me.

NIGEL: Can you afford 60 quid, like?

ROGER: I'm a party animal! You should know that by now, Nigel! But yeah, I manage to save out of my salary – but not much. What they pay here! It's just beer money. But wouldn't it be great to look back in two years, say, and to give the finger to Management!? Peggy [the portering manager] said to me the other day that my job here was 'the real thing' and should be more important to me than the band! What?! Is she kidding? Has she never had *fun*? [We laugh] These people have never had fun. In fact, only the porters round here have a sense of humour. Don't you think? Like, I was messing round outside the Theatre – me and Roy McMadden having a bit of a laugh with the paper bins – and this guy comes out of the Theatre and says, 'This is an Operating Theatre!' [Roger mimics a serious, slightly English, officious voice]

NIGEL: A doctor?

ROGER: Aye, probably a doctor. They act like automatons: 'Do this, do that' ... But I like portering: it's a good job. I've applied for permanent porters' jobs in X-ray, Physio, Specimens, and some others. Specimens is okay 'cos you're your own boss: no one tells you what to do when. But I don't want to work shifts. Just nine-to-five. And I also don't want to work in the Theatres. Too hot and too boring.

NIGEL: Like today!

ROGER: Aye: back in Theatre! I was in ICU on Friday. Then I went to Endoscopy, then upstairs to ICU again and then back. But they're *all* deadly: you sit in an office even hotter than the buckie and do nothing: just wait for patients to come out of Theatre.

NIGEL: How many go in and out?

ROGER: I think four went in this morning. But I'm fed up; surrounded by old men – fifty year olds – who just do crosswords all the time. This afternoon I'm gonna steal the radio and play some Black Sabbath! I did it before 'cos they were playing an Andrew Lloyd Webber tape five times over! So I swopped it for some Sabbath – and that's what the patients heard when they came out of surgery!

NIGEL: Cool! What did people say?

ROGER: No one said anything or complained; I mean, I didn't have it on too loud ... But Saturday night was great! At the Uni bar. The bassist from our band knew the DJ and he played the first track on *our* cd and he liked it!! And other folks said they did too; and these two girls came up and said 'Was I the singer?' and then they hugged me! And I thought if this is what it's like then *great*!... [We laugh] I think our band is destiny, like! It's amazing how things have all come together. So something good must come out of it! But if I make it big I don't want riches or fame. I don't want a big mansion; just a normal house.

NIGEL: Why?

ROGER: I don't want to be mobbed in the street so I can't go anywhere. But, still, when I got hugged by those girls I thought: I could get used to this! [We laugh] I mean, I'd never preach or go to church – I can't understand all these preachers on street corners saying you're all going to Hell. I can't stand that. I remember this preacher on TV saying he'd not let his grandchildren go to a Black Sabbath concert and how Ozzie Osborne was going to Hell unless he repented. And then Ozzie was asked what he made of that, and Ozzie's dead cool and he just says people take everything so literally and they shouldn't. Music is just about having fun! Don't get me wrong: I've got respect for everything. But I'd also do anything now: no holding back. I'm totally mad.

NIGEL: Sure!

ROGER: And no one is like us, with our lyrics. Right?! [I laugh]

NIGEL: Where did that song you showed me come from: 'Two-Faced Son of a Bitch'?

ROGER: I wasn't thinking of anyone in particular when I wrote it; it just came out. There's all this stuff about 'love' and about how the world is a great kind of place; but it's not! So say it ... Anyway, I got shouted at for taking too long a lunch-break yesterday: Dick, the 'boss', the gaffer, the oldest, said a patient was sitting in the corridor waiting for me ... [We move to exit the canteen] Alright Big Joe? [Roger greets another porter in operating theatre greens]

NIGEL: Who's that?

ROGER: Big Joe from Endoscopy. I teach his kid karate: he's very keen. But I'm a big softie really: the thing is to entertain them as well as teach them ... I know loads of people working here. Most are friends of friends, like.

> A few are from school, like the two lasses in X-ray. Two lookers! Aye: there are great women around here in the hospital. I'll have to get myself one! [We laugh. As we leave the canteen Roger accompanies our departure with a loud, heavy-metal rendition on an imaginary air-guitar]

How does Roger appear in this extract? A young man finding his way into an environment of adult work. He still lives at home with his parents, and so can spend most of his earnings on himself recreationally. His energy is that of a young man, perhaps, too: little sleep, many 'extra-curricular' activities, from pubbing to singing with his band (recording and writing songs) and idolizing Ozzy Osborne and Black Sabbath, to practising and teaching karate, to fancying pretty girls. But I want to reach beyond the stereotypical. I was struck by a phrase Roger used in talking about his music, his love of it and his hopes for it. In music 'there are no limits', he told me, and, 'I have no limits'. It is in terms of the limits and limitations of the work environment at Constance, and those who seem to thrive within it, that Roger's essential character stands out, for me. The porters are distinct but the remainder of the employees, and the hospital regime with its institutionalized practices, do not know what to make of Roger, he feels: his energy, his liveliness – his wildness – his joy, his involvements, and fascinations. He is surrounded by officiousness and bossiness, by people who act as boring and old automata who cannot esteem the real thing in life and who have no *fun*.

In the terms I have been promoting, Roger finds a life-force within himself that for different reasons and in different ways kicks against the threatening fixities and narrowness he finds around him. A mobility perspective – the recognition, after Bateson, that things accrue identities by their movement relative to other things from a point of perception, or, after Woolf, that consciousness is fed by a constant movement of 'atoms' of events to which it attends – calls upon the analyst to attend to the way movement constitutes our human being-in-the-world. Here is Roger, then, whose identity accrues from the moving array of 'others' that he brings within his purview: his nights out, his drinking and spending money, his friends, his band, his music, 'his' Black Sabbath, his partying, his singing, his hugging girls, his enthusiasms, his karate, his fearlessness, 'his' Easterneuk – and also 'his' Constance Hospital. The hospital is somewhere where he has many acquaintances, somewhere where he has applied for various long-term jobs, somewhere where he plays his music, somewhere where he is bored, somewhere where he argues with management about the meaning of 'fun' and a place with bosses and shift-work that he disdains. In short, the induction that Roger and I shared on our first few days working together at Constance is followed by Roger *initiating himself* into this environment by having conscious relations with a moving array of people, locations, settings, events, issues, practices, rules, and futures. Likewise, Roger moves Constance Hospital into relation with all that exists beyond it – before it and around it – in his life. He contextualizes Constance Hospital within an extensive and expanding environment of personal landmarks through which he moves in his conscious awareness of himself – and his expression of that awareness.

When I say that Roger 'kicks against' fixities at Constance that might constrain him, there is also a literalness to this: as I would accompany him on occasion through the extensive corridors and stairwells of the site (when our work trajectories momentarily overlapped), Roger would show me his karate prowess by kicking his way through the innumerable fire-doors in our path and also kick down signs pinned high up on noticeboards we might pass. Roger's movement through Constance left behind a wake of noise and events that proved his course. But then, what of those fixities? Roger expressed himself surprised at how his fellow workers set themselves limits and appeared to accept the terms and institutional norms of the hospital. But how does a mobility perspective account for the way that Roger *did* accommodate himself to the work hours of a job at Constance Hospital, the tedious ways in which those hours had to be passed, to the uniform that must be worn, to a management with the power to hire and sack, to allocate work-roles and overtime, and even presume to counsel him on the necessities and values of life? (His reputation for idiosyncrasy among other staff, including porters, often had him classed as a fool, if not mentally backward.) Here, too, I can say, movement is key to our understanding. Roger did not control the impressions that others had of him and the ways they acted towards him – the ways *others* may have moved 'Roger Weir' relative to their points of perception – insofar as he and the movements of *his* conscious being existed alongside a plurality of other things-in-the-world, inorganic and organic – people and materiel, human institutions and norms, natural processes and human traditions – some of them engaged in the movement of their life as is Roger. To return to a phrasing of Bateson's (1973, 426), we might describe Roger Weir as existing as an 'organism-in-its-environment'; he is discrete in the journey of which his life comprises and in the meanings and identities he construes on his moving way, *but he is not alone in the world.* As an individual-in-its-environment Roger Weir is embarked upon a distinct voyage of activity-in-the-world, of movement and sense-making in the world, but he is not alone in this and is surrounded by other conscious beings doing likewise, amid settings of natural, social, and cultural processes.

We might say that the 'fixities' facing Roger Weir at Constance Hospital translate in a mobility perspective into an appreciation of two very different things. On the one hand there is what might be described as the 'structural power' of natural, social and cultural processes: the way that nature, society and culture – body, social relationships, institutional norms – may constrain the forms of an individual life. On the other hand there is the 'existential power' that human beings possess and by which they give their lives movement, direction, meaning, and purpose (Rapport 2003). Residing within the individual body, 'existential power' defines force, the will, the energy, in a word the agency, whereby human beings move so as to produce effects in their worlds – indeed, effect worlds. Roger Weir's life at Constance Hospital is an amalgam of the structural power of an institution that would fix him within certain workaday norms alongside the existential power of an individual (Roger) embarked on a life course of meaning-making, and alongside the existential power of other human individuals acting likewise.

EXISTENTIAL POWER AND SOCIAL STRUCTURES

The concept of existential power points to human beings as discrete centres of energy (Bateson's individual 'energy sources'). Existential power is something metabolic, pertaining to individuals as embodied physical organisms. It is also something intelligent, pertaining to the capacity to sense and make sense. While human bodies are permeable, and while each is dependent on energy transfers across these borders, each is also the site, throughout its life, of an inescapable physical and experiential distinctiveness – a distinct site of perception and interpretation – which separates it and distinguishes it from the rest of the world. As discrete centres of energy, individuals begin, from before birth, to become distinctly themselves: to accrete identities and personalities. This takes place through activity-in-the-world: through movement and through assessment of what the senses relay to be the results of that movement. Not only does the energy behind this activity-in-the-world remain individually based, it is also individually directed. From the moment the individual energy source begins moving in its environment and becoming itself, a unique history of embodiment, of worldly engagement, unfolds and grows and that encompasses its own logics, its own habits, its own ways of doing and being, its own purposes. What grows is a personal, environing 'sensorium' in which human minds dwell: their 'phenomenological subjectivity' as individual actors (Fernandez 1992, 127, 134–5). Since each individual centre-of-energy is driven by its own metabolism, within its own embodiment and along its own historical course of activity-in-the-world, how each will react to other things is not determinable. It is difficult, for instance, if not impossible, to predict how one human being will affect another human being with whom it comes into contact. This is so for three reasons: firstly, to repeat, because each is set upon its own life course, each is engaged in furthering a life-world whose direction and logic has been distinct from the moment 'it' began. Secondly, because each engages with others from the position of outsider, each is dependent on bodily sense-making apparatuses that are discrete and distinctive to each human being, imbuing the latter with its own perspective on the world and no other. Thirdly, because the sense-making procedures of each are characterized by a creativity – a 'randomness' even (Rapport 2001) – which makes their generation of movement and meaning unpredictable even to themselves.

Human social life, according to Ralph Waldo Emerson (1981, 278), should be conceived of as a mix of 'the two elements, power and form': a meeting between individuals' 'native force' and social conventions (between the existential and the structural). A 'prosperous' social life, Emerson concluded, depends on how 'soundly' or 'sweetly' individuals were able to effect the meeting between their inner force and the surface conventions that act as both buffers and points of contact between themselves and others.

'We live amid surfaces, and the true art of life is to skate well on them' (Emerson 1981, 275). Where does this leave Roger Weir? His 'skating well' over the surfaces of institutionalism and hierarchy at Constance Hospital depended on an ability to interpret those norms, ranks and locations in ways that brought him satisfaction, even when the resources available to gain satisfaction were not equally distributed. Others are better placed to make the hospital work for them and use the structural power of a hierarchy of ranks and a tradition of local practices towards their own ends. Roger's existential power to move himself relative to a world of otherness and to make sense – including Constance Hospital, with its norms of behaviour, ranks and locations – comes up against the existential power of others at the hospital to do likewise. How well, then, did Roger skate? At the very least, a mobility perspective ensures that proper respect is given to Roger's capacity and, indeed, nature for being responsible for the ways in which the exterior social forms of the hospital entered into his consciousness and were processed into sense. He was, in my view, determined to hold his own at Constance, to keep moving. As he would advise me: 'Keep on rockin', Nigel!'

Is this too glib? (When my fieldwork came to an end, Roger remained at Constance: this can be construed as both a victory and a defeat.) To engage with the 'quintessential' (and polythetic) phenomenon of movement is not to fail to see factors inhibiting such movement: social structures, hierarchies and borders, normative, institutional, and communitarian. Scaling up analytically from Roger Weir to include others and populations of human others, what does a mobility perspective reveal concerning human beings moving in social-structural spaces? The story remains a human one. Changes in scale do not alter the logic by which we are to understand the human story of bodies in motion. The ground of human being and of the analysis of such human being is the specific consciousness enjoyed by different individual actors at different moments of their lives: consciousness that owes its character and its contents to the movements each engages in. Human intentionality, human agency, lodged in individual moving bodies, provide the key causal factors. There is the social structure and symbolic system by which social and cultural life is formally practised, and there is the diversity and particularity of how those structures and symbols are interpreted and experienced by individuals in the moments of their lives.

The effects of non-human 'actors', from storms to computers, plants to pets, depend on how these are attended to, interacted with and moved in relation to, by human actors.

There is, then, no separate domain of 'the social' that runs according to its own kinds of impersonal facts, forces, laws, and energies (Rapport 2009). The social structures that impinge upon individuals' movement do so by virtue of how they are worked by other human individuals. The power we accord to these structures, the effects 'they' have, exist by virtue of the sense made of them by individuals who assume particular positions in relation to them – individual movement gives them meaning. Movement and the knowledge that derives from it remains a human attribute. Social structures may 'possess' inertia, a fixity, and objectivity that human actors cannot fail to attend to, but their creation, their animation (working) and their destruction derive from and continue to depend on, their (individual) human operatives.

This is not to deny the different positionings relative to those social structures that different individuals and populations of individuals – different 'operatives' of those structures – will adopt. Roger Weir is a porter at Constance: he is not a doctor or administrator or nurse. Equally the members of Tamil, Roma, Zanzibari, Tongan, Mambwe and Ju|'hoansi San populations on which this volume has focused find themselves and the movements they undertake objectively situated relative to certain social structures according to the categories they are seen to occupy. Adopting a mobility perspective is to insist, nevertheless, on not accepting such structures and categories as things-in-themselves. To tell the story of these populations and of Roger Weir alike is to tell of those moments when individual human beings move themselves relative to aspects of their environments that *they* perceive: from storms to computers, plants and pets, from social and cultural forms to natural facts to other human beings, from social structures to systems of symbols.

ENVOI

It is fitting we take our leave of Roger Weir with him explaining to me his latest impressions of his workplace:

ROGER: I did the Main Theatre on Thursday and Friday and Saturday and Sunday, Nigel! Really boring. Alone in the buckie there. No real work all day – till you clean the theatre at the end ... So I went to Peggy on Monday and I said I didn't wanna be there in Theatre always. And she says how they're training me there in case an opening arises in the future. And I say I've got my band to consider. And she says: 'This hospital should be your first priority 'cos the band does not earn me a living' ... What? Yeah right!! And this place is a heap of shit. So many people are horrid. You ask them a question and they look at you like you've lost your marbles. I'm gonna start telling people straight what I think of them. Sod it!

NIGEL: But the porters are okay.

ROGER: Yeah! Not the porters. They're great. I love them. But they can't 'take' me in Theatre – I'm too hyper ... Being alone in the buckie outside the theatre just depresses me. I need company: people around me, before too long. Also: the small space; I feel cooped up, claustrophobic ... I also had a hepatitis shot. 'Cos I'm fed up with the jobs I've been given, cleaning blood and who knows what off the floors. I could get a superbug. I'm fed up playing by the rules: I feel like there's one set of rules for some and one for the rest of us ... *And* I don't know the shifts I'm on – and I need to know that with these gigs coming up for the band ... But I'd do the gigs whatever the consequences.

Bibliography

Auge, Marc. 1995. *Non-Places: Introduction to an Anthropology of Supermodernity*. Translated by J. Howe. London: Verso.

Bateson, Gregory. 1936. *Naven*. Cambridge: Cambridge University Press.

——1973. *Steps to an Ecology of Mind*. London: Intertext.

——1980. *Mind and Nature: A necessary Unity*. Glasgow: Fontana.

Berger, John. 1984. *And our Faces, my Heart, Brief as Photos*. London: Writers and Readers.

Carter, Paul. 1992. *Living in a New Country: History, Travelling and Language*. London: Faber & Faber.

Clifford, James. 1986. 'Introduction: Partial Truths'. In *Writing Culture*, edited by George Marcus and James Clifford, 1–26. Berkeley: California University Press.

Cohen, Anthony P. 1992. 'Self-Conscious Anthropology'. In *Anthropology and Autobiography*, edited by Judith Okely and Hellen Callaway, 221–41. London: Routledge.
Dalsgaard, Steffen, and Morten Nielsen (eds). 2015. *Time and the Field*. Oxford: Berghahn.
Emerson, Ralph Waldo. 1981. *The Portable Emerson*. Edited by C. Bode. Harmondsworth: Penguin.
Fernandez, James. 1992. 'What it is Like to be a Banzie: On Sharing the Experience of an Equatorial Microcosm'. In *On Sharing Religious Experience: Possibilities of Interfaith Mutuality*, edited by Jerald D. Gort, Hendrik M. Vroom, Rein Fernhout and Anton Wessels, 125–35. Amsterdam: Rodopi.
Hart, Keith. 1990. 'Swimming Into the Human Current'. *Cambridge Anthropology*, 14 (3): 3–10.
Jackson, Michael. 2002. *The Politics of Storytelling*. Copenhagen: Museum Tusculanum Press.
——2015. *Harmattan*. New York: Columbia University Press.
Leach, Edmund. 1977. *Custom, Law and Terrorist Violence*. Edinburgh: Edinburgh University Press.
Lévi-Strauss, Claude. 1975. *The Raw and the Cooked*. New York: Harper Colophon.
——1981. *Tristes Tropiques*. New York: Atheneum.
Nielsen, Morten (ed.) forthcoming. *Urban Times*, *Ethnos* special issue.
Rapport, Nigel. 1993. *Diverse World-Views in an English Village*. Edinburgh: Edinburgh University Press.
——1994. '"Busted for Hash": Common Catchwords and Individual Identities in a Canadian City'. In *Urban Lives: Fragmentation and Resistance*, edited by V. Amit-Talai and H. Lustiger-Thaler, 129–57. Toronto: McClelland and Stewart.
——1997. *Transcendent Individual: Towards a Literary and Liberal Anthropology*. London: Routledge.
——2001. 'Random Mind: Towards an Appreciation of Openness in Individual, Society and Anthropology'. *The Australian Journal of Anthropology*, 12 (2): 190–220.
——2003. *I am Dynamite: An Alternative Anthropology of Power*. London: Routledge.
——2008. 'Gratuitousness: Notes towards an Anthropology of Interiority'. *The Australian Journal of Anthropology*, 19 (3): 331–49.
——2009. 'Power and Identity'. *The Handbook of Power*. Edited by S. Clegg and M. Haugaard. London: Sage.
——2014. 'Time'. In *Social and Cultural Anthropology: The Key Concepts*, 416–21. London: Routledge.
Rapport, Nigel, and Andrew Dawson (eds). 1998. *Migrants of Identity: Perceptions of Home in a World of Movement*. Oxford: Berg.

Ruesch, Jurgen, and Gregory Bateson. 1951. *Communication: The Social Matrix of Psychiatry.* New York: Norton.

Simmel, Georg. 1971. *On Individuality and Social Forms.* Edited by D. Levine. Chicago: Chicago University Press.

Sørensen, Ninna Nyberg, and Karen Fog Olwig (eds). 2003. *Work and Migration: Life and Livelihoods in a Globalizing World.* London: Routledge.

Woolf, Virginia. 1938. *The Common Reader.* Harmondsworth: Penguin.

Notes on Contributors

ADA I. ENGEBRIGTSEN is Research Professor at Norwegian Social Research, Centre for Welfare and Labour Research, the University College of Oslo. Her main research topics are minorities, interethnic relations, mobility, Roma issues, family and children. She has published internationally on all these issues.

ØIVIND FUGLERUD is Professor of Social Anthropology at the Museum of Cultural History, University of Oslo. His research interests include nationalism, diaspora formations, politics of cultural representation and aesthetics. He has published a number of works on the conflict in Sri Lanka and its consequences, including *Life on the Outside: The Tamil Diaspora and Long-Distance Nationalism* (Pluto Press, 1999). In 2015, he, together with Leon Wainwright, published the edited volume *Objects and Imagination, in Perspectives on Materialization and Meaning* (Berghahn, 2015).

JENNIFER HAYS is Professor of Social Anthropology at the Institute for Archaeology and Social Anthropology, the Arctic University of Norway (Tromsø). She has been working with the San, nomadic hunters and gatherers in southern Africa, since 1998. She has published numerous articles and edited volumes on educational and human rights issues as they affect this marginalized indigenous population in Namibia, Botswana and South Africa. Her recent book, *Owners of Learning: The Nyae Nyae Village Schools over 25 years* (Basler Afrika Bibliographien, 2016) explores the educational challenges facing the highly mobile Ju|'hoansi living in the Kalahari Desert of north-eastern Namibia.

KJERSTI LARSEN holds a PhD in Social Anthropology from University of Oslo, Norway. She is a professor in the Department of Ethnography, Numismatics and Classical Archaeology, at the Museum of Cultural History, University of Oslo, where she has also been head of department (2003–6). She was also Adjunct Professor in the Department of

International Development & Environmental Studies, University of Life Sciences, Norway (2001–8). She was a member of the ICARDA Board of Trustees (2003–9). She has been a visiting scholar at the Centre for Cross-Cultural Research on Women, University of Oxford; Centre d'Études Africaines, École des Hautes Études en Sciences Sociales; International Institute for the Study of Islam in the Modern World (ISIM), Leiden University, Netherlands. She conducts research in Muslim societies in East Africa, on the Swahili Coast, particularly on Zanzibar (1984–present), and in Northern Sudan, mostly in the Bayoda desert (1997–2008). She has several international publications, including the monograph *Where Humans and Spirits Meet: The Politics of Rituals and Identified Spirits in Zanzibar* (Berghahn, 2008) and the edited volume *Knowledge, Renewal and Religion* (Nordiska Afrikainstituttet, 2009). Her main interests include ritual and performance; religion, knowledge and gender; everyday-life politics and social change; identity, mobility and connectivity in African societies and the Indian Ocean region; and Islam, orthopraxy and mundane Muslim practices.

VELINA NINKOVA is a PhD candidate in Social Anthropology at the Institute of Archaeology and Social Anthropology, at UiT, the Arctic University of Norway. She has worked with San communities in Namibia and Botswana since 2008. Her current research focuses on the challenges former hunter-gatherers face with formal education. Other fields of interest include indigenous peoples-state relationships, anthropology of childhood, visual anthropology. She is the author of the ethnographic film *Children of the Kalahari*, which explores the everyday realities of Ju|'hoan children on a resettlement farm.

ARNE ALEKSEJ PERMINOW is Associate Professor of Social Anthropology and keeper of the Oceania Collection at the Museum of Cultural History, University of Oslo. His research interests are environmental knowledge, ritual aesthetics, and social values and its changes through movements in time and space. He has published works on the relationship between everyday experience, ritual aesthetics and sociality based on multiple fieldworks on Kotu Island in Ha'apai, Tonga, including *Captain Cook and the Roots of Precedence in Tonga: 'Leading' and 'Following' as Naturalised Concepts*

(History and Anthropology, 2001) and *Food Presentations Moving Overseas: Ritual Aesthetics and Everyday Sociality in Tonga and Among Tongan Migrants* (in Ø. Fuglerud and I. Wainwright, *Objects and Imagination. Perspectives on Materialization and Meaning*, Berghahn Books, 2015). He has curated several exhibitions, including *Startpaths across the Pacific, Narratives of Origin in Oceania* (Museum of Cultural History 2006).

NIGEL RAPPORT is Professor of Anthropological and Philosophical Studies in the Department of Social Anthropology at the University of St Andrews, and Head of the School of Philosophy, Social Anthropology, Film and Music. He is the founding director of the Centre for Cosmopolitan Studies. His research interests include social theory, phenomenology, individuality, literary anthropology, symbolic interactionism, cosmopolitanism, liberalism, and freedom. Recent books include *Anyone, the Cosmopolitan Subject of Anthropology* (Berghahn, 2012) and *Distortion and Love: An Anthropological Reading of the Art and Life of Stanley Spencer* (Ashgate, 2016).

JAN KETIL SIMONSEN is an Associate Professor and former head (2004–8) of the Department of Social Anthropology, Norwegian University of Science and Technology. He is the co-editor of the *Norwegian Journal of Anthropology*. His research interests include mobility and migration, kinship, ritual studies, childhood studies, and visual anthropology. His latest book is *Beyond the visual: Sound and image in ethnographic and documentary film* (Intervention Press, 2010, co-edited with Gunnar Iversen).

Index